Joseph Cook

Conscience

With preludes on current events

Joseph Cook

Conscience
With preludes on current events

ISBN/EAN: 9783337418502

Printed in Europe, USA, Canada, Australia, Japan

Cover: Foto ©berggeist007 / pixelio.de

More available books at **www.hansebooks.com**

BOSTON MONDAY LECTURES.

CONSCIENCE,

WITH PRELUDES ON CURRENT EVENTS.

BY JOSEPH COOK.

"Ethical science now teaches not so much that man has conscience, as that conscience has man." — DORNER.

DeLoss M. Tompkins.

BOSTON:
HOUGHTON, OSGOOD AND COMPANY.
The Riverside Press, Cambridge.
1879.

COPYRIGHT, 1878,
BY JOSEPH COOK.
All rights reserved.

*Franklin Press:
Stereotyped and Printed by
Rand, Avery, & Co.,
Boston.*

INTRODUCTION.

THE object of the Boston Monday Lectures is to present the results of the freshest German, English, and American scholarship on the more important and difficult topics concerning the relation of Religion and Science.

They were begun in the Meionaon in 1875; and the audiences, gathered at noon on Mondays, were of such size as to need to be transferred to Park-street Church in October, 1876, and thence to Tremont Temple, which was often more than full during the winter of 1876-77, and in that of 1877-78.

The audiences contained large numbers of ministers, teachers, and other educated men.

The thirty-five lectures given in 1876-77 were reported in the Boston Daily Advertiser, by Mr. J. E. Bacon, stenographer; and most of them were republished in full in New York and London. They are contained in the first, second, and third volumes of "Boston Monday Lectures," entitled "Biology," "Transcendentalism," and "Orthodoxy."

The lectures on Biology oppose the materialistic, and not the theistic, theory of evolution.

The lectures on Transcendentalism and Orthodoxy contain a discussion of the views of Theodore Parker.

The thirty lectures given in 1877-78 were reported by Mr. Bacon, for the Advertiser, and republished in full in New York and London. They are contained in the fourth, fifth, and sixth volumes of "Boston Monday Lectures," entitled "Conscience," "Heredity," and "Marriage."

In the present volume some of the salient points are: —

1. The definition of conscience as "that which perceives and feels rightness and obligatoriness in choices" (p. 17).

2. A fuller definition (p. 25), together with a distinction between what conscience includes and what it implies (pp. 25–28).

3. A study (in Lectures II. and III.) of the relations of ethical and biological science, or of the effect of the approval and disapproval of conscience upon the countenance and gesture.

4. A reply (in Lectures IV., V., and VI.) to the agnosticism of Matthew Arnold.

5. A criticism (in Lecture VI.) of the positions of Mansel as to the definitions of the infinite and absolute.

6. A consideration (in Lectures VII. and VIII.) of conscience as the foundation of the religion of science.

7. A series (in Lectures IX. and X.) of literary illustrations of conscience from Victor Hugo and Shakspeare.

The theory as to conscience advanced in these lectures is in general accord with the ethical school represented by Kant, Dugald Stuart, Price and Edwards, and, among later German writers, by Lotze, Wutke, Hofmann, Ulrici, and Rothe. It emphasizes that view of the moral faculty which materialism opposes, but which is admitted to have had, from Plato's time to the present, the greatest number of scholarly adherents. The peculiarity of the volume is in the use made of the most recent biological science.

A consideration of the origin of conscience will be found in the lectures on Heredity.

The committee having charge of the Boston Monday Lectures for the coming year consists of the following gentlemen: —

His Excellency A. H. RICE, Governor of Massachusetts.
Hon. WILLIAM CLAFLIN, Ex-Governor of Massachusetts.
Prof. E. P. GOULD, Newton Theological Institution.
Rev. WILLIAM M. BAKER, D.D.
Rev. WILLIAM F. WARREN, D.D., Boston University.
Prof. L. T. TOWNSEND, Boston University.
E. M. MCPHERSON.
ROBERT GILCHRIST.
Prof. GEORGE Z. GRAY, D.D., Episcopal Theological School, Cambridge.

Prof. EDWARDS A. PARK, D.D., Andover Theological Seminary.
Right Rev. BISHOP PADDOCK.
Prof. E. N. HORSFORD.
Hon. ALPHEUS HARDY.
Rev. J. L. WITHROW, D.D.
A. BRONSON ALCOTT.
RUSSELL STURGIS, Jr.
Right Rev. BISHOP FOSTER.
REUBEN CROOKE.
SAMUEL JOHNSON.
WILLIAM B. MERRILL.
Prof. B. P. BOWNE.
M. R. DEMING, *Secretary.*
B. W. WILLIAMS, *Financial Agent.*

HENRY F. DURANT, *Chairman.*

PUBLISHERS' NOTE.

IN the careful reports of Mr. Cook's Lectures printed in the Boston Daily Advertiser, were included by the stenographer sundry expressions (applause, &c.) indicating the immediate and varying impressions with which the Lectures were received. Though these reports have been thoroughly revised by the author, the publishers have thought it advisable to retain these expressions. Mr. Cook's audiences included, in large numbers, representatives of the broadest scholarship, the profoundest philosophy, the acutest scientific research, and generally of the finest intellectual culture, of Boston and New England; and it has seemed admissible to allow the larger assembly to which these Lectures are now addressed to know how they were received by such audiences as those to which they were originally delivered.

CONTENTS.

LECTURES.

		PAGE
I.	Unexplored Remainders in Conscience	3
II.	Solar Self-Culture	35
III.	The Physical Tangibleness of the Moral Law	61
IV.	Matthew Arnold's Views on Conscience	87
V.	Organic Instincts in Conscience	117
VI.	The First Cause as Personal	143
VII.	Is Conscience Infallible?	171
VIII.	Conscience as the Foundation of the Religion of Science	201
IX.	The Laughter of the Soul at Itself	229
X.	Shakspeare on Conscience	255

PRELUDES.

		PAGE
I.	Insurrections of Hunger	5
II.	Bachelor and Family Wages	35
III.	English Precedents in Civil-Service Reform	61
IV.	The Duties of Opulence to Missions	87
V.	Enfranchised Ignorance in the South	117
VI.	Indigent Infidelity	143
VII.	California as the Door to China	171
VIII.	Free Tabernacles in Great Towns	201
IX.	Magdalen in Cities	229
X.	Young Men in Politics	255

I.

UNEXPLORED REMAINDERS IN CONSCIENCE.

THE EIGHTY-FIRST LECTURE IN THE BOSTON MONDAY LECTURESHIP, DELIVERED IN TREMONT TEMPLE, OCT. 1.

Zwei Dinge erfüllen das Gemüth mit immer neuer und zunehmender Bewunderung und Ehrfurcht, je öfter und anhaltender sich das Nachdenken damit beschäftigt: der bestirnte Himmel über mir, und das moralische Gesetz in mir.—KANT: *Sämmtliche Werke*, ed. Hartenstein, v. 167.

Kant's "two things that strike me dumb;"—these are perceptible at Königsberg in Prussia, or at Charing Cross in London. And all eyes shall yet see them better; and the heroic Few, who are the salt of the earth, shall at length see them well.—CARLYLE: *Shooting Niagara: and after?* vi.

CONSCIENCE.

I.

UNEXPLORED REMAINDERS IN CON-SCIENCE.

PRELUDE ON CURRENT EVENTS.

In the year 1877 America has seen her first, but probably not her last, insurrection of hunger. Low-paid labor has at least occasionally not had enough to eat; and therefore a thin flame of fire burst out of the hitherto rarely ruptured social soil on a line extending from Baltimore to San Francisco. This ominous, wavering, but intense radiance rose from a fruitful, a largely unoccupied, and a monumentally unoppressed country. Our cities gather to themselves the tramps, the roughs, and the sneaks; several of them contain organized bands of emigrant communists; and this loose material caught fire when the sudden flame shot up from the volcanic crevice. We were not very swift in putting down the conflagration. It happens, therefore, that in a land which has twice been washed in blood, and was an hundred years old, society suffered painfully for several weeks, from a wide-spread strike of railway laborers, a riot of roughs and sneaks, and an inefficient self-defence. We are all agreed that it takes two to make a bargain; and even low-paid labor occasionally forgot that first principle of social science. The chief trouble came, however, not from the workingmen, and not from the real princes of capital, but from second-rate business managers, who hardly know how to make a fortune except by cut-throat competition.

How many railways of this country are in receivers' hands? We talk of various cures for the ills of our railway strikes; but is not one of the most practical remedies a requisition by law that every railway corporation, and every moneyed company that is in debt and yet in receivers' hands and in business, shall be compelled to lay aside at least one per cent of its income as a sinking fund to pay its debt? We must in some way insist upon it, that unprincipled competition shall not grind the faces of the poor. Your Vanderbilt did not grind those faces. I do not know that Thomas Scott did; however, I think he is paid a large salary not for his knowledge of legitimate railroading, but for his knowledge of illegitimate railroading. No railway deserves to succeed whose managers would tremble if their ledgers were turned inside out and read by the whole American people. Here, for instance, are two railway companies, each containing a dozen men. A majority in each company secretly arrive at an understanding with each other. They form in fact, though not in name, a third company. That third collection of managers owns no railroads; but it has a majority in two companies that do own, perhaps, competing lines. By making a ring, they can turn aside, for a time, to their own uses, a very large part of the profits of both these railway companies. The conspirators have not a wheel, they have not a track, of their own; but they put into their pockets a lion's share of the proceeds of the companies in which they have a majority. They place profits on board one car, and turn this off upon a side track; and, when the train of their enterprise reaches the station farther on, they announce that there is nothing left for the stockholders; and of course, if stockholders suffer, workingmen must.

Mines and factories and railways are likely to be heard of in the maturity of the American republic, not as loudly, but perhaps as pointedly, as the cotton-field and the rice-swamp were in its infancy. As the Old World has had peril enough from industrial questions to make already classic much of the literature of the conflict between labor and capital, this New and young World does not act unwisely in turning attention, with all the power of American conscientiousness and shrewdness, upon the inquiry, What are comfortable wages, and how can they be paid? Is it possible to arrive at a definition of starvation wages?

Suppose that a man were to put forward the proposition that any thing less than twice the cost of the uncooked food for a family containing several small children is starvation wages to the unassisted father of that family, would you think such a position very heretical? Regard for a moment the perplexities of low-paid labor. After all, the pulpit has the right, and the platform, — especially if it be as free as this one, — at least this will take the privilege of looking into the vexed arithmetic of the very poor. A man has in his family a wife and three children. He must therefore feed five mouths. What do you pay for your board each week? Five dollars, perhaps, and it is not very good at that. What could you get the bare food for, without any charges for cooking or rent? Three dollars? Two and a half? Two? I should not like to live and do hard work ten hours a day on food that cost less than two hundred cents a week, or twenty-nine cents a day. You would not. But I am at the head of a family; and my wife has only health enough to cook the food, and take care of the children and the house. She really earns nothing except in acting as a housekeeper and as a mother to my children, — there are three of them, — and now I must maintain five persons. Food certainly cannot keep soul and body together, and cost less on the average than a dollar a week. I must starve, or have five dollars a week for the uncooked food of my family. How much do I earn a day? A dollar, without board. My children cannot earn any thing. If I obtain work every day, I have at the end of the week a dollar left to pay for rent and every thing else. Is it hard times with my family? The children must have shoes, or they will be hooted at in the street when they go to the public school. America is, indeed, kind. She opens the school to the poor. But I ought to be able to put shoes on the feet of my children; and yet I cannot always put coats on their backs, nor even can I have ragged calico for my babes at times, for I have but a dollar a day, and they can earn nothing, and my wife is a little ill. But I must send my children to school, or I drop to a lower social scale. My children ought to go to church, but they have nothing to wear. I ought to send my wife to church; I ought to go myself; and I am not to be excused for keeping away, because it would be better for me if conscientiousness were diffused throughout the community, and I know that one great object of the church is to diffuse conscientiousness, in order that property may be safely diffused. I ought to be, with my brethren of the laboring class, in God's house every sabbath day; and I ought to be there with my children. But I must pay five dollars a week for the food of my family; and I earn but a dollar a day or a little more,— some of my brethren earn but ninety cents,— and I work but six days in the week. I want to get my children a few school-books. I ought to take a newspaper. There must be now and then a doctor's bill paid. I must have a little coal in the winter; and it is not possible for me to buy it as the millionnaire does, in great quantites: I must buy it by the basket, and my wood in little parcels. And it is hard times. I have just been dropped from employment. There is often not much for me to do. I cannot always find work six days of the week.

Undoubtedly there are some corporations that have paid as wages more than they have received as profits. Workingmen have occasionally been retained in place at a temporary loss to their employers. But supply and demand are the law of business, and I am discussing the dull average sky of low-paid labor under that rule, and not the starry exceptions.

I sat in a parlor beyond the Mississippi, with two leaders of business, one of them a millionnaire, and the other nearly such, and we added up the necessary expenses of a family of five, in which children are supposed to be too young to labor remuneratively; and we found that such a family could not very well live through a year respectably in our climate, and according to the standard of the workingmen of America, if the father is their only support, and is paid less than ten or twelve dollars a week. The low-paid laborer often has wages that are less than six hundred dollars a year. Your Massachusetts Bureau of Labor in 1875 published a large collection of details from the life of families in this Commonwealth, and asserted that "the fact stands out plainly, that the recipient of a yearly wage of less than six hundred dollars must get in debt." (*Pub. Doc.* No. 31, 1875, p. 380.) I know how high wages often are in the ranks of skilled labor; but, as John Bright used to say, "the nation lives in the cottage." I undertake to maintain here in Boston, where heresies are popular, the astounding proposition, that if the unassisted father of a family of three children who cannot labor remuneratively is paid no more than twice the cost of the unprepared food for his family, he is on starvation wages.

THE LECTURE.

When Samuel Taylor Coleridge, the poet, was a poor boy and a charity-scholar in London, he was one day walking along the Strand, at an hour during which the streets were crowded, and was throwing out his arms vigorously toward the right and the left. One of his hands came into contact with a gentleman's waistcoat-pocket; and the man immediately accused the boy of thievish intentions. "No," said Coleridge, "I am not intending to pick your pocket. I am swimming the Hellespont. This morning in school I read the story of Hero and Leander, and I am now imitating the latter as he swims from Asia to Europe." The gentleman was so much impressed by the vividness of the imagination of the lad, that he subscribed for Coleridge's admission to a public library, which began the poet's education. The beginning of all clearness on the multiplex topic of Conscience is to make a distinction between picking a pocket and swimming the Hellespont. [Applause.] The external act may be precisely the same, although the inner intentions differ by celestial diameters. It is natural to man, however he obtained the capacity, to make a distinction between meaning right and meaning wrong. Not only did this gentleman and the poet-boy not stop on the Strand to settle the question whether the intuitional or the associational theory in ethics is correct, but the urchin, coasting down the long mall of Boston Common, would not stop for that purpose, were he struck by some careless coachman with the lash. He would look up, and immediately ask, "Did you mean to do that?" And if he sees that it was the result of accident, he excuses the coachman; but if he finds that the coachman meant mischief, he accuses him accordingly. Horace represents the children's games at Rome, as proceeding according to the laws of conscience: —

> "Pueri ludentes, Rex eris, aiunt
> Si recte facies."
>
> *Epist.*, lib. i. Ep. i. 59.

Just so the babe that cannot speak, building its card-house on your parlor-carpet, will look up when you trample down its castle, and ask, not verbally, but by action, whether you meant to do that; and if it ascertains that you did not, you will be excused; but if you intended to destroy the work of the babe, that untutored human constitution will re-act against you. This babe, building its card-castle, has not been evolved very far in human experience. It has not had a long time in which to develop, by considering questions of utility, a tendency to notice the difference between meaning right and meaning wrong, and to make a distinction between the outward act and the inner intention. However it arises, whether according to the theory of Herbert Spencer or Alexander Bain and others of their school, whom I imagine sitting yonder on my left, or according to the theory of Kant and Rothe and their followers, whom I imagine sitting there on my right, we have here and now, as human beings, a tendency to ask whether any one who injures us means to do so, or does so accidentally; and according to the intention we judge the external act. In one case it is picking the pocket: in the other it is swimming the Hellespont. [Applause.]

There are two schools represented by these stately auditors of ours, invisible but tangible here; and when I turn to Spencer and Bain on my left, I find conscience called fallible, educable, vacillating. John Foster, in a celebrated essay, says that, among human spiritual possessions, there is nothing so absurd and chimerical as conscience. It is a bundle of habits. Pascal affirms that " conscience is one thing north of the Pyrenees, and another south." We have a fifth listener here, Dean Mansel, a pupil of Sir William Hamilton, and who built on the only boggy acre of his master's generally sound territories. Even he asks incredulously how conscience obtains the right to rule the other faculties. (MANSEL, *Limits of Religious Thought*.) But if I turn to Immanuel Kant, I find him uttering the amazing proposition, that " an erring conscience is a chimera." There is no such thing. (*Tugendlehre*, ix. 38.) I ask Rothe yonder what he says about that statement, and he bows assent to the whole of it. (*Theol. Ethick*, ii. 29.) I cross the German Sea to Scotland, and enter the parlor of Professor Calderwood, teacher of ethics in the university at Edinburgh, where Sir William Hamilton taught, and that scholar is putting Kant's proposition, that an erring conscience is a chimera, into the foreground of his last work. (*Handbook of Moral Philosophy*, p. 81.) Fichte supposed himself to have annihilated the doctrine that there can be any such thing as an erring conscience. (*Sittenlehre*, iv. 227.) Stuart Mill sits yonder, and Rothe here looks Stuart Mill in the eyes; and as I gaze into their faces, I do not find that Rothe and Fichte and Kant are as likely to be looked out of countenance as Mill and Spencer.

Nevertheless there must be some way of explaining the difference between these honest men. We have the same debates among ourselves. We are accustomed to affirm that conscience has something divine in it; and that which is divine does not mislead us, does it? But we say also that conscience is not infallible; it is erring. The Bible itself speaks of conscience as seared, blunted, and blinded. We have Scriptural warrant for saying that the conscience may be seared as with a hot iron. And yet the Bible does speak of a Light that lighteth every man that cometh into the world, and that in the beginning was with God, and was God. Can that be seared with a hot iron? Can God be blinded? Plainly there are two doctrines in the Scriptures on this subject, or rather two points of view. These opposing schools are not defending propositions that really contradict each other. They stand at different points of vision; and so the different popular ideas concerning conscience are apparently self-contradictions, because we do not notice that they are taken up from opposing outlooks.

Whenever you find yourself in a mental fog, attend to the duty of definition.

What is conscience? It was my fortune to spend the first three months after the close of three years' theological study, alone on Andover hill, with the use there of the best theological library in New England. I had had the usual professional instruction in religious science; but to my humiliation I must confess that when I asked myself what I meant by conscience, it was impossible for me to give a distinct definition. Rothe, in the last edition of his occasionally eccentric but really great work on Theological Ethics (sect. 177, anm. 3), carries his disaffection with the term conscience so far as to exclude it from his volumes altogether as scientifically inadmissible and devoid of accurately determined logical contents. I had been authorized to teach such as were foolish enough to listen a few propositions concerning religious truth; but I could not define conscience. I set myself to work, and it was nine days before any adequate light dawned upon that point. What I am now to put before you I have often tested by putting it before scholars, and I do not know that an essential syllable of it has ever failed to receive indorsement. Nevertheless I ask no man to adopt my theory of the moral sense; I am speaking here, as always, not to scholars, and not to teachers of religious science who honor us with their presence, but to the average inquirer; to the person who, beginning to think for himself, finds that he must, first of all, learn how to think, and that, on many a great topic, he needs to know what has survived in the struggles of scholars with each other, age after age, and to know this from men who have time to examine the record.

1. Conscience, according to the loose popular idea of it, is the soul's sense of right and wrong.

2. Conscience, according to the strict scholarly idea of it, is the soul's sense of right and wrong in its moral motives, that is, in its choices and intentions.

3. On the one hand, it is clear that conscience, defined in the loose popular way, as only the sense of right and wrong, implicitly includes the action of the judgment as well as of the moral perceptions and feelings.

4. Since judgment is fallible, conscience, defined as a spiritual multiplex, including both the moral sense and the judgment, is fallible, and may justly be spoken of as often blinded, erring, and seared.

5. A still greater fault in the loose popular definition is that it makes no explicit distinction between the outer act and the inner intention.

6. The conscience, according to this definition, is supposed to be a compound of faculties by which we decide on what is called the rightness or wrongness of external acts, and as such is, of course, doubly fallible, and may with scientific justice be pronounced erring, vacillating, and often self-contradictory.

7. On the other hand, if conscience be defined in the strict scholarly way, as the soul's sense of right

and wrong in the sphere of its own intentions, the judgment or purely intellectual activity of the soul is distinguished from the moral perception and feelings, and, therefore, in this definition, does not constitute a fallible factor in conscience.

8. A man does infallibly know whether he means right or wrong in any deliberate choice.

9. If, therefore, conscience be supposed to be, as the strict definition describes it, the soul's sense of right and wrong in its own choices and intentions, and in those only, conscience is infallible within its field.

10. In this sense and in that field, conscience is not educable.

11. It follows from this definition, that right and wrong, strictly understood, belong only to choices and to intentions as including choices. "Nothing," says Kant, "can possibly be conceived in the world, or even out of it, which can be called good, without qualification, except a good will." (*Grundlegung*, Sect. 1.)

12. External acts, taken wholly apart from the intentions which led to them, have only expediency or inexpediency, usefulness or harmfulness; and their character in these respects is ascertained by the judgment and not by the conscience.

13. If, however, we employ the loose definitions of conscience, there is an important distinction to be made between absolute and relative right. Absolute right is the conformity of the action of a free moral agent to the fitness of things as they are; relative

right is the conformity of our choices and intentions to the fitness of things, as, with the best light within our reach, we believe them to be. Conscience points out to a man the relative right; that is, the good and evil in his intentions.

14. When we are about to form an intention, conscience looks forward, and perceives its character; when we form it, conscience points out its nature as good or bad; after we have executed it, conscience does the same.

15. Conscience is thus antecedent, concurrent, and subsequent, in relation to every act of choice.

16. It is demonstrable that before and without the verdict of the judgment as to what the external results of an intended act will be, conscience approves or condemns the intention as in its own nature good or bad.

17. All languages make a distinction, as does conscience, between the right and the expedient, the right and the prudent, the right and the advantageous.

18. But conscience not only perceives the difference between a good intention and a bad: it feels that the good intention ought, and that the bad ought not, to be adopted and carried out.

19. It is demonstrable that conscience inflicts remorse for an evil choice taken alone, or for a bad intention formed but not carried out.

20. Every intention has two sides, — rightness or its opposite, and obligatoriness or its opposite.

21. The former distinction is perceived, the latter felt.

22. Conscience, therefore, may be briefly and provisionally defined as a faculty including both a perception and a feeling, — *a perception of right and wrong in the nature of choices and intentions; and a feeling that right ought, and wrong ought not, to be carried out by the will. Conscience is that which perceives and feels rightness and obligatoriness in choices.*

Such is the definition with which we set out on a course of thought in which it is hoped there may be discussed John Stuart Mill's views, Herbert Spencer's, Matthew Arnold's, as well as Kant's and Rothe's and Butler's, or the entire conflict between the associational and the intuitional, and between the latter and the pantheistic theory, concerning the loftiest of the faculties possessed by man. At the close of the enlargements and verifications of these propositions which are to come in subsequent lectures, there will be inferences of a sort which I hope will do something to blanch the cheeks of unscientific thoughtlessness. Everywhere we are to proceed according to the principles of inductive science. We are to ask, What are the facts in man's inmost life, and what its relations to the nature of things? We are to infer from uncontroverted facts concerning the moral sense, what its nature is. We are to judge it by its effects. I am not asking you, in any thing I have thus far put before you, to accept Mill's theory, or Rothe's, Herbert Spencer's, or Kant's. I am asserting here and now only that a distinction is to be made between external acts and inner intentions, and that the peculiar prerogative of conscience is to tell us what is right or wrong within the sphere of intentions.

You notice that I have admitted the propriety of all our popular, and of course of all the Scriptural language, concerning the possibility that the conscience may be seared with a hot iron; but I insist also that there is in us an original capacity to judge of the difference between right and wrong intentions, and that as clearly as we see that the whole is greater than a part, we see that meaning right is something different from meaning wrong.

There are ethical axioms, as there are mathematical axioms; and if exact research establishes axioms in ethics, you will know how to build on them after the pattern shown in the Mount. In the mysteries of man's moral nature there is a Mount that burneth yet as with fire, and that cannot be touched, and which, if we could see it in its unexplored remainders, we should ask to have screened from us, for no man ever passed forty days and forty nights there without coming down with such a glory on his face as to need a veil.

In spite of the distinctions which I have indicated, you say that it is not clear that judgment is not concerned in determining whether a motive or intention is right or wrong. When I was in Syria, I saw many strange fruits, and could occasionally pluck down a pomegranate, and look at it, weigh it in my hand, notice its subtle fragrance, and finally taste it. Now, no doubt the intellectual faculties do pluck down motives from the tree Igdrasil, and no doubt we stand as lawyers before the court of conscience, and make pleas, often very mischievous ones. It is beyond controversy that the judgment is a fallible faculty, and that I do weigh the Igdrasil pomegranate in the intellectual hand, and that it does bring the fruit to the lips; but it is only the tongue that tastes the pomegranate. By an intellectual act, we bring the motive clearly before conscience, and conscience perceives its flavor. It is not the fingers that taste the strange fruit. The eyes know nothing of flavor. There is no sense possessed by man by which the flavor of the pomegranate can be ascertained, except that which rests in the tongue. Without the sense of taste, there is no perception of flavors; without conscience, there is no perception of the difference between right and wrong. Neither in the former nor in the latter case can perception be acquired. A being without conscience, however highly endowed intellectually, cannot be taught to feel the distinction between what ought to be and what ought not to be. We do not reason with the Corliss engine, to teach it that it should plunge its pistons regularly.

We can imagine a being possessed of the intellectual equipment of the Aristotles and Bacons, or the executive ability of the Napoleons and Cæsars, and yet without a perception of the difference between right and wrong. *We can picture to ourselves a creature possessed of that perception and yet without any feeling, when right has been seen, that it ought to be followed; but neither popular nor scientific language would permit us to say that such a being has a conscience.* This crucial fact shows that the moral sense must be made to include both a perception and a feeling; but the latter may be weak, and conscience yet exist.

I define conscience as that within us which not only perceives what is right in moral motives, but also feels that what is right ought to be chosen by the will. You may be puzzled by the question whether conscience is not sometimes inoperative or dead. I know that this *feeling* that what is right ought to be followed, may have greater or less force; but the *perception* that there is a distinction between right and wrong in intentions, or between meaning to do well and meaning to do ill, I hold is clear in every man down to the limits of sanity; and that, although the magnetic needle may not always be followed, although the crew may be crazy and not look at the card, there is in the needle a power that makes it point to the north whenever it is balanced on a hair point, and allowed to move without fetters.

A man does infallibly know whether he means to be mean or not, and he does infallibly feel mean whenever he means to be mean.

We are so made that the distinction between right and wrong in the sphere of intentions is as evident to us in moral action as the superiority in size of a whole over a part is in the sphere of mathematics.

I beg Mr. Mill's pardon: I am not using the word intuitive, which he dislikes and which Kant honors. Here and now I insist on nothing more than the proposition that self-evident truths are the basis of mathematics, and that self-evident truths are the basis of ethics, and that we perceive all such truths directly. They are matters of supreme certainty. There is a difference between the right hand and the left in the soul's choices among moral motives, and men are as sure concerning that as they are concerning the proposition that every change must have an adequate cause. Distinguish, then, between the fingers that pluck down the fruit, or the intellectual faculties that discuss intentions, and the peculiar sense that tastes them. [Applause.] I may almost define conscience as the tongue that tastes the flavor of intentions. [Applause.]

Conscience is an original faculty, although in activity it draws the other faculties of the soul into its service. Taste, or the power of perceiving the beautiful, is an original faculty; but it uses all the other faculties. So memory is an original faculty; but its activity implies the action of many other faculties. The power to appreciate the ludicrous is an original faculty; and its activity, like that of conscience, implies the exercise of both perception and feeling. There is no more reason for calling conscience a merely composite power, or simply the entire list of human faculties applied to moral truth, than for calling taste a composite power, or simply the entire list of the faculties applied to the laws of beauty. At the last analysis of taste and memory, and the power to perceive the ludicrous, each is found to have a separate peculiar function of its own; and so has the moral faculty.

Moral discernment differs from merely intellectual discernment in that the former is, and the latter is not, necessarily followed by a feeling of obligation. The discernment of the ludicrous differs from merely intellectual perception in that the former is, and the latter is not, necessarily followed by the feeling which prompts to laughter. A similar contrast exists between the perception of the beautiful and merely intellectual perception. There is a region of the soul in which perception and emotion appear to be inseparably blended, and to constitute one faculty, as two elements unite in water, which is yet but one substance. This region, as Sir William Hamilton has remarked, yet needs a nomenclature.

Whatever the origin of our powers of taste and memory and wit, each is here in human nature; and so is conscience. As Sidgwick has remarked (*Methods of Ethics*, London, 1877, second ed. p. vi.), the teacher of ethics is no more called on to investigate at the outset of his discussions the origin of conscience than the geometer to investigate the origin of those perceptions as to space and time upon which geometry and arithmetic are built. Under the topic of Heredity (see vol. v. of the *Boston Monday Lectures*), it is my purpose to discuss the origin of conscience; but here and now I exhibit the faculty only as an inalienable portion of balanced human nature, like memory or the perception of the ludicrous, or the sense of the beautiful.

The chief advances of science have come from the study of unexplored remainders. We have in conscience a perception of the distinction between right and wrong in moral motives. But what lies behind that perception? The difference exists in the nature of things, apparently. But what is meant by the nature of things? There is in conscience a feeling that we ought to follow what we perceive to be a right moral motive, and ought not to follow what we perceive to be a bad one. But what lies behind the terrific weight of the word *ought?*

Take the single syllable *ought*, and weigh it, my surprising sceptical friends, and do so according to the sternest rules of the scientific method. How are we to ascertain what this word weighs, unless it be by experiment? What experiment shall we try with it, if it be not that of weighing over against it something very heavy? What shall we weigh against the one word ought? Here is a soldier with an empty sleeve. There was a day when the question arose, whether he ought to go to the front in the war. He had to maintain father and mother; and the word home is supposed to be a very weighty one. Heavier than the word father or mother is the word wife. He weighed that word and the others with it against the one word ought; and father and mother and wife went up in the scale, and *ought* went down, and he went to the front. Is *ought* scientifically known to weigh any thing? Here is another soldier who had father, mother, wife, and children, to weigh against that insignificant syllable; and he weighed them, in the mornings and the noons — in both the sacred twilights, as they say in India —

and in the midnights. Father, mother, wife, and children were words to which he allowed their full weight. He was the only support of his family, but the one word *ought* again and again carried up the weight of these weightiest contradicting syllables. What if this soldier and that could have put into the left-hand scale all that men value in wealth and honor or reputation? I will not suppose the word honor to have any other meaning than reputation, for I cannot weigh ought against ought; and a man ought to maintain his honor. We must not be so unscientific as to weigh a thing against itself. But we put in here, outward standing among men, and wealth, and life. If you please, sum up the globes as so much silver and the suns as so much gold, and cast the hosts of heaven as diamonds on a necklace, into one scale, and if there is not in it any part of the word ought — if ought is absent in the one scale, and present in the other — up will go your scale laden with the universe, as a crackling paper scroll is carried aloft in a conflagration ascending toward the stars. [Applause.] Is it not both a curious and an appalling fact, this weight of the word ought — and yet a fact absolutely undeniable? Where is the materialist or the pantheist who dares assert that I am making this syllable too heavy? You may weigh against that word every thing but God, and it will outweigh all but himself. I cannot imagine God weighed against ought. Precisely here is the explanation of a mystery. God is in the word ought, and therefore it outweighs all but God. [Applause.] There is your first unexplored remainder.

But, my friends, we must be analytical in order to be brief.

I. Conscience in full activity includes, —

(1) A direct perception of right and wrong in choices.

(2) A feeling that right ought and that wrong ought not to be performed.

(3) Complacency in the right, and displacency in the wrong.

(4) A sense of personal merit in the performance of the right, and of personal demerit in the performance of the wrong.

(5) A delight or pain, bliss or remorse, according as the choices are right or wrong.

(6) A prophetic anticipation of reward for the performance of right, and of punishment for the performance of wrong.

The fundamental proof that conscience in full activity exhibits the six special methods of action here named is to be found in accurate observation of what takes place in our own mental and moral experience.

II. An important distinction exists between what conscience includes and what it implies.

(1) A direct perception of the freedom of the will is not one of the activities of conscience; but the fact of such freedom is a necessary inference by a single step of reasoning from the sense which conscience gives us of personal merit and demerit; for it is self-evident that these can be the qualities of only voluntary action.

(2) A direct perception of the fact of the Divine

existence is not one of the activities of conscience; but the fact is a necessary inference by a single step of reasoning from the perception of a moral law and the sense of obligation to it included in conscience. The moral law, of which the existence is proclaimed in the very structure of conscience, and so is spiritually tangible by conscience, is an Eternal Somewhat not ourselves which makes for righteousness. But the Plan in that Somewhat is a thought, and there cannot be thought without a thinker; and so the Somewhat, in all the high activities of conscience in connection with the intellectual faculty, is recognized as a Some One. This recognition is so necessary and universal that the fact of the Divine existence has often been called a strictly intuitive truth, and the assertion made that conscience and the soul's consciousness of God are one.

(3) A direct perception of the fact that a future state of personal existence awaits man is not one of the activities of conscience, but is an inference from the prophetic anticipations irresistibly asserting themselves in conscience, that reward and punishment await him beyond death; and also, according to Kant, from the demand which conscience makes for the soul's absolute perfection, and the practically necessary condition of a duration adequate to the complete fulfilment of the moral law.

It is well known that Kant makes the freedom of the will, the fact of the divine existence, and that of immortality, postulates, that is, presuppositions, of conscience, and asserts that "the truth of these ideas no sophistry will ever wrest from the conviction of even the commonest man." (*Dialectic of Pure Practical Reason,* vi.)

III. The effects of conscience arise both from what it includes and from what it implies.

Among the effects resulting from both these sources are : —

(1) A sense of an approval or disapproval from a Divine Somewhat or Some One not ourselves, according as we are influenced by good or bad intentions.

(2) A bliss or a pain, each capable of being, at its height, the acutest known to the soul; the former arising when what ought to be has been done, and the latter when what ought not; and the two alternating or acquiring final permanence according as our approval or disapproval of ourselves, and our feeling of our approval or disapproval by a Divine Somewhat or Some One not ourselves, alternate or acquire final permanence.

(3) A prophetic anticipation that both our approval and disapproval by ourselves and by a Divine Somewhat or Some One not ourselves are to continue beyond death, and to have consequences affecting us there as personal existences.

(4) An authority, imperativeness, and inner necessity, arising from a source in us, and yet not of us, and against which, in the activities of conscience, the will and all the human faculties are utterly powerless.

In these three propositions and their subdivisions,

I venture to summarize my definition of conscience. If we put into the definition of the moral sense not only all it includes, but all it implies, we overload the definition, and accurate psychological observation will not justify our analysis. This is the fault of many mystical definitions. On the other hand, if, in our description of conscience, we do not take into view what it implies, as well as what it includes, our account of the moral sense is not true to the facts of life: it is cold, inadequate, and palpably unscientific. This is the fault of many rationalistic descriptions.

The novel point in the definition and description of conscience here attempted is the distinction between what conscience includes and what it implies. The activities of conscience and the effects of conscience are to be distinguished from each other in that the former contain only what the organic actions of the faculty include, while the latter result from both what those actions include and what they imply.

Only he who takes into view both what the activities of the moral faculty include, and what they imply, can have any proper conception of the awe and mystery and might of conscience.

In the preliminary definition I have used the word *sense;* for that may mean either a *perception* or *feeling*, and conscience includes both a perception of rightness, and a feeling of oughtness. This latter word is in standard use in the Scottish philosophy. "It is not plainer," said Richard Price (*Review*, chap. 6), "that figure implies something figured, solidity resistance, or an effect a cause, than it is that rightness implies oughtness." (See also CALDERWOOD, *Handbook of Moral Science*.) Butler taught that so far from conscience being a perception or a feeling alone, "it probably includes both." I am aware how much I venture in giving a definition of a term as to the full meaning of which there is up to this hour only too little agreement among experts. (See HOFMANN, *Das Gewissen*, Leipzig, 1866; the best recent German work on Conscience.)

From the dawn of ethical investigation, fragments of the definition of conscience now given have been appearing, although they have rarely been combined into a self-consistent whole.

Butler confines the action of conscience to the sphere of intentions: "Will and design constitute the very nature of actions as such, and they are the object, and the only one, of the approving and disapproving faculty." (*On the Nature of Virtue*, Diss. II.) He describes, though he does not discuss, its prophetic office: "Conscience without being consulted, without being advised with, magistérially exerts itself, and if not forcibly stopped naturally and always of course goes on to anticipate a higher and more effectual sentence, which shall hereafter second and affirm its own." (*On Human Nature*, Sermon II.)

The most elaborate recent treatise in German on conscience defines it as "a fixed readiness (coming into activity with inner necessity in a given act of will) to institute a comparison between the given act of will and a law as standard in the same instant with the act of will, touching us from outside ourselves, and unconditionally claiming for itself authority." (HOFMANN, *Das Gewissen*, p. 83.)

Here and now I use the numbered propositions of this discussion only as the outline which this lecture is intended to draw in bold contours; and I leave you to take the point of view of practical philosophy, without asking you to decide to-day between the Mills and the Spencers on the one hand and the Kants and the Rothes on the other. These two sets of listeners will indorse these propositions as statements true to human nature. There is within us the power of perceiving the difference between right and wrong in the sphere of intentions. We have a feeling that the right ought to be followed, and that the wrong ought not to be. We have a sense of merit and demerit, or of approval and of disapproval of ourselves. Our instincts assure us that there is an approval or disapproval above our own. We have a bliss or pain, according as we feel this approval or disapproval from ourselves, and from Somewhat or Some One not ourselves. Lastly, there is in conscience a prophetic office, by which we anticipate that consequences, closely concerning us as conscious personal existences, will follow us beyond death. In all these particulars conscience acts without the consent of the will. It puts forth its activities by a mysterious inner necessity, which although in us is not of us. It claims for itself, therefore, in the constitution of man unconditional supremacy. "Had it strength as it had right," says Bishop Butler, "had it power as it had manifest authority, it would absolutely govern the world." (*On Human Nature*, Sermon II.)

I defy any student of the laws of the human soul as recorded in the unpartisan record of the languages and literatures of all the nations, or any man who will be faithful to the scientific method in the introspective study of his own experience, or any candid and clear thinker, to deny, in the name of inductive science, the existence in the moral faculty of either of the seven traits here ascribed to it.

Think of the unexplored remainders beyond each one of the ascertained scientific facts concerning Conscience. Where is the seat of that Authority which speaks in the mysterious but wholly undeniable weight of the word ought? Where now is He who is the Light that lighteth every man that cometh into the world, and that in the beginning was with God and was God? There are men who do not perceive the absolutely unfathomable glory of Christianity either as a philosophy or as a life, and who ask vaguely where He is who spoke once as never man spake, and since has governed the centuries? Where is He whose pierced right hand lifted heathenism off its hinges, and turned into another channel the dolorous and accursed ages? To me, too, on humble and struggling paths in the valleys of thought, as well as to your Kants and your Rothes, aloft there where the sky-kissed peaks of research gaze upon the coming sun, the sublimest as well as the most organizing and redemptive truth of exact ethical science is the identity of the moral law and the Divine Nature. Wherever the moral law acts, there Christianity finds the personal omnipresence of Him whom we dare not name,— Father, Son, and Holy Ghost; Creator, Redeemer, Sanctifier; One God, who was, and is, and is to come. At this miraculous hour, the Light that lighteth every man that cometh into the world is, not was. It is scientifically known that this Light has its temple in Conscience. But it has been proclaimed for ages by Christianity that God is One, and that our Lord is as personally present in every breath of the Holy Spirit in the latest days, as he was in that breath which he breathed on his disciples when he said, "Receive ye the Holy Ghost." Our cheeks may well grow white and the blood of the ages leap with a new inspiration, when, standing between Christianity and science, we find the thunders of the one and the whispers of the other uttering the same truth. It is a familiar doctrine to Christianity, that our bodies are the temple of Somewhat and Some One not ourselves. That Some One Christianity does, although physical science does not, know by an Incommunicable Name. There are connections between religion and science here of the most overawing moment; and in the whole field of the truth concerning Conscience they are the vastest unexplored remainders. [Applause.]

II.

SOLAR SELF-CULTURE.

THE EIGHTY-SECOND LECTURE IN THE BOSTON MONDAY
LECTURESHIP, DELIVERED IN TREMONT TEMPLE,
OCT. 8.

Das Gesicht eines Menschen sagt in der Regel mehr und interessanteres als sein Mund. Auch spricht der Mund nur Gedanken eines Menschen, das Gesicht einen Gedanken der Natur aus.—
SCHOPENHAUER: *Parerga und Paralepomena*, ii. 509.

Ben discerneva in lor la testa bionda;
Ma nella facce l'occhio si smarria,
Come virtu ch'a troppo si confonda.
DANTE: *Purgatorio*, viii. 34.

II.

SOLAR SELF-CULTURE.

PRELUDE ON CURRENT EVENTS.

WHOEVER becomes an incendiary or an assassin in the conflict between labor and capital, let him suffer the full penalty of the law, whether he be a millionnaire or a Molly Maguire. [Applause.] The riffraff rioter, the petroleum Communist, the fire-bottle loafer, are enemies of the human race; and if they defy the law, a republic must treat them with that kind of mercy which Napoleon showed toward the original Communists of Paris, when he closed the French Revolution by a whiff of grape-shot. Asked to account for the splintering of the Church of St. Roche, he said, "It is false that we fired first with blank charge: it had been a waste of life to do that." As a republican of the American, and not of the Red or Communistic species, I passed in Paris some thankful moments, leaning against the rabbets and plinths of St. Roche Church, which show splintered by that shot to this hour.

The American lower ranks contain three different sets of men — the unenterprising, the unfortunate, and the unprincipled. The shiftlessness of the unenterprising sometimes needs the spur of hunger. It is good political action, as well as good morals, to insist that if any man will not work, neither shall he eat. The unfortunate who are not unprincipled will not long remain unfortunate. Our civilization, therefore, will need to concern itself chiefly with those who are really without aspiration or principle enough to occupy their opportunities of rising in American society. Hampden and Cromwell, Adams and Washington, have made it possible for any one to rise in the United States who has the strength and the will to do so. If any one does not rise, it must be because he lacks either energy or principle. "I began with twenty-five cents," said a millionaire to his discontented workingmen on the Mississippi last summer, "and every one of you has the same opportunity." That was a distinctively American speech. Commonly the cripples and the roughs, the very unfortunate and the utterly unprincipled, are at the bottom of society in democratic great cities.

The lower classes abroad are composed very differently from the American. Dives and Lazarus in the Old World have, and here they have not, hereditary positions. The mobility of our society is such that Dives or his sons may sink to the position of Lazarus, and Lazarus or his sons may rise to the position of Dives. We have no law of primogeniture. We have no inherited or artificial social rank. The sons of the poor and the rich easily change positions. It follows, from this fact, that the cause of the rich man in America is every man's cause. A man is a man, even if his father was rich. But it follows also, from the same fact, that the cause of the poor man is every man's cause. A man is a man, even if his father was poor.

"For a' that, and a' that,
A man's a man for a' that."

But if a man will work, shall he eat? There is a distinction to be made between family wages and bachelor wages. At the bottom of the collisions of labor and capital, which caused ten cities in America to listen not long ago to volleys of sharp shot, was the competition of bachelor wages with family wages. John here has a dollar and a half a day, and can barely support his family. James yonder has no work, and is a bachelor, and, of course, is willing to labor for eighty cents a day. Vanderbilt says he could have manned all his railways by paying only eighty cents a day for labor. James comes to John, and says, "There is a strike, and you are unwilling to labor for eighty cents a day; but I am willing, and will take your place." John replies, "James, if you do that, I will kill you." James says, "If you shoot me, the soldiers will shoot you." John answers, "I will stop the trains, and you shall not run them." He is as good as his word. That is what it is: a conflict between bachelor wages and family wages. The soldiers appear when the roughs and the sneaks begin to fire round-houses and trains. The workingmen did not intend to burn up valuables; but they meant to keep ill-paid labor from outbidding them in a competition which was reducing wages. Their mode of doing this was to stop railway traffic, — no doubt a most suicidal as well as criminal procedure. Low-paid labor forgot two things: first, that it takes two to make a bargain; and, second, that it is not one of the rights of labor to prevent labor.

When two representatives of the workingmen — family wages on one side and bachelor wages on the other — come thus into collision in the youth, or rather in the infancy of the Republic, the sign is talkative about much yet to come in the maturity of our land. There is a hope possessed by many, that the collisions between capital and labor may, in America, be settled by reason, and not by force, and settled, not according to the ideas of capital on the one hand, nor according to those of labor on the other. Force, in the riots of the Communists in Paris, settled the question for a while on the side of petroleum roughs and sneaks, or the unprincipled portion of the lower classes. In some other parts of Europe, hereditary position and wealth and absolute government have settled the question with equal injustice by force, although with less noise, and on the side of capital. If, in America, this question can have fair discussion from the friends of both labor and capital; if, as is perhaps not easily possible, the question can be kept out of the hands of political demagogism [applause]; if it can be lifted up early to a plane of thought substantially Christian, — then America, in settling the question for herself, will assuredly help much to settle it for the world.

> "When a deed is done for freedom, through the broad earth's aching breast
> Runs a thrill of joy prophetic, trembling on from east to west;
> And the slave, where'er he cowers, feels the soul within him climb
> To the awful verge of manhood, as the energy sublime
> Of a century bursts full-blossomed on the thorny stem of time."
> LOWELL : *The Present Crisis.*

One full blossom has appeared on the American branch of the tree Igdrasil, in the abolition of slavery; perhaps an hundred years hence the time will be ripe for the appearing of another blossom in the peaceable settlement of the conflicts between labor and capital. [Applause.]

But when John refuses to allow James to take his place at eighty cents a day, John has his children in mind. What are comfortable wages? If starvation wages were correctly defined in a previous discussion, shall we not ask, with sharp attention, What are natural or just wages? My proposition, which I do not ask any one to defend, is that just wages will not violate the rights of children and of old age. By this I mean that whoever is willing to labor physically the legal number of hours a day should be paid enough to insure him, if he is prudent and economical, and has no bad habits, a living for himself and his children while they are too young to labor remuneratively, opportunity to educate them, and some support for himself and wife when the power to labor shall have ceased. That is only enough to give the State the strength of its citizens. That is only enough to make firm the ground-sill which must lie under what your Wendell Phillips calls the heavy working of republican institutions. If public sentiment, if arbitrating boards, if friends of capital or labor, will turn attention upon the facts officially ascertained and published by your Massachusetts Bureau, it will be found that children's rights are deeply complicated with this whole question of wages. Why, you have in this Commonwealth now 104,000 illiterates out of a population of 1,600,000. Twelve thousand of these illiterates are native born. More than ninety thousand are foreign born. But, whether born here or abroad, they failed to learn to read and write, chiefly because it was necessary for them to assist in the support of their families. It is understood very well by all who have looked into the statistics on this question, that "children under fifteen years of age" — I am reading the very words of the report of 1875 of the Massachusetts Bureau of Labor — "supply by their labor from one-eighth to one-sixth of the total family earnings of the wage class in this Commonwealth. On children, parents depend for from one-fourth to one-third of the entire family earnings." Families with most children occupy usually the worst tenements. Without children's earnings a majority of the 397 families, which your Bureau visited in 1875, would have fallen into poverty or debt. With the assistance of children, there was only in a few cases a possibility of a family acquiring a competence, that is, of having a home of its own, even after the father was sixty years old.

Now, I am drawing near a time of life when I ought to begin to think of founding a home; and it would certainly seem to be hardship to me, if at forty or forty-five years of age I could not have a little place that I could call mine. But what if at fifty or sixty I could not? How do I know that a grandson of mine will not be a laborer by the day? How do you know that the haughty children of Boston may not have grandchildren that will not be haughty on a dollar a day? We in this country are all members of each other, for there is no hereditary position for any man; and what if at sixty your descendant could not, if industrious, economical, and without bad habits, have a little home of his own when his power to labor ceases? I conversed with a celebrated manufacturer in old Manchester, England, once, and put the question to him, whether any large percentage of the ordinary unskilled operative class in England could have homes of their own. "Not three in a thousand," said he. Although you may not believe it until you examine the facts, low-paid labor in this country, in the very language of your own Massachusetts Bureau, "has only in a few cases a possibility of acquiring a competence," that is, a home of its own.

My proposition is simply that we must not violate children's rights, and add thus to the strangely crescent class of the illiterate; and that we must not violate the rights of people in extreme age, and thus fill up the ranks of the poorhouse and of all who depend on charity. We are closing the youth of the American republic. We are drawing near its majority, and children's rights and the rights of age we must protect by a proper consideration of wages.

Who is to arbitrate? Mr. Mundella says that arbitration abroad has effected more than any thing else to heal difficulties between workingmen and their employers. I fear greatly that the State of Ohio, in now leading off American sentiment in the direction of governmental interference between capital and labor, is taking a somewhat unadvised step. This whole matter is going into politics. Demagogues are to discuss it. We are to have all kinds of deformers mingled with reformers on this theme. If I venture much in introducing it here, I do so because I believe that only a diffusion of conscientiousness by the churches, only the bringing of all classes, rich and poor, into those relations which belong to them when they are on the floor of God's house and measured by His standards, can ultimately give safety to society under republican institutions. [Applause.] You have no better arbitrating board in America between labor and capital than the voluntary system in the American church. Give us a glorious American church, and we will settle for the world here, and settle peaceably, the conflicts which only the bayonet has been able to put down abroad.

You Christianize Magdalen, you wish to Christianize Lazarus, you would Christianize Dives. Has not the hour come in America when religion, in the name of political science and of Him who once had not where to lay His head, should stand erect in her shining garments, and teach, in no craven or apologetic tone, not only that republican institutions must Christianize Magdalen and Lazarus and Dives, but that, first of all governmental institutions in the world, they must Christianize Cæsar? [Applause.]

THE LECTURE.

Dante, describing the angels whom he met in the Paradiso, impresses us at once with their external glory and their spiritual effulgence. Invariably he makes the former a result of the latter. With closer faithfulness to physical science than he dreamed, and building better than he knew, he sings: —

> "Another of those splendors
> Approached me, and its will to pleasure me
> It signified by brightening outwardly,
> As one delighted to do good;
> Became a thing transplendent in my sight,
> As a fine ruby smitten by the sun."
> *Paradiso*, canto ix. 13-19.

Dante says of Beatrice, as he saw her in the Paradiso, that, —

> "She smiled so joyously,
> That God seemed in her countenance to rejoice."
> *Paradiso*, canto xxvii. 105.

Allow me to adopt this last line of Dante's, and all it suggests, as a description of what I mean by solar light in the face of man. This radiance ought to be by us, as it is by natural law, most searchingly distinguished from all lesser illuminations. *Its specific difference from every other light is that in it God seems to overawe beholders and to rejoice.* It is scientifically incontrovertible that there is sometimes seen such a light in the present world. Many a poet and seer and martyr and reformer, or woman of the finest fibre, has at times had a face that has looked like porcelain with a light behind it. But this is not solar light, unless it have in it that specific overawing difference which Dante names. The mysteriously commanding and glad light is to be distinguished from merely æsthetic or intellectual luminousness in the countenance, by a peculiar moral authority, incisive regnancy, and unforced elateness, bliss, and awe. The radiance cannot be counterfeited. It can come into existence only on inexorable conditions. The appearance and disappearance of the solar light in the face of man are governed by fixed natural laws. Is it possible to discover any of them?

First of all, I ask you to look at the whole topic of solar self-culture through the lenses of the coolest inductive research. Put aside all mysticism; fasten the attention only on visible facts, as well known to be a part of human experience as that men walk or breathe; build only on the granite of the scientific method, and let us see what structure can be erected by the use of blocks cut on a line with the natural cleavage of the rock from this unhewn quarry, that is, by untutored, indisputable propositions certified by daily observation.

1. There is sometimes in the face a solar look.

2. There is sometimes in the face an earthy look.

3. The former arises from the activity of the higher nature when conscience is supreme.

4. The latter arises from the activity of the lower nature when conscience is not supreme.

5. The earthy look, other things being equal, quails before the solar look.

6. The merely intellectual light in a face quails before the solar light when other things are equal.

7. Merely æsthetic light, or that arising from the action of the faculties addressed by what is commonly called culture, quails, other things being equal, before the solar light.

8. The light of merely executive force, other things being equal, quails also.

9. The intellectual, the æsthetic, the executive, and all other light combined, quail, other things being equal, before the solar light.

10. It follows necessarily that only such self-culture as brings this light to the face can give its possessor all the power possible to man.

11. Only such self-culture can cause the lower forms of culture to stand in awe before it.

12. The only complete and the only victorious self-culture, therefore, is scientifically known to be solar self-culture.

Be Greeks, gentlemen, long enough to believe that every change, and therefore the variation in the inner illumination of the countenance, must have an adequate cause. How is it that this peculiar commanding light springs up from within the multiplex whole of our physical organism? Your materialist will say that certain emotions increase the tension of the mechanism of the eye; and that, therefore, external light is more readily reflected by it, and that we have hence, apparently, a new light in the eye when those emotions are active. But what is to be said of the light that beams from the forehead, and from the cheeks, and seems to be capable of beaming from the whole exterior of our mysterious form? That radiance does beam from the forehead; it does beam from the cheeks; and why might it not, if this capacity of the organism to shine were once put into full action, beam from the whole man? The materialist would say that the particles of matter in the cellular integument are capable of re-arrangement by certain emotions, and that they reflect light better on account of this re-arrangement. But what gives those emotions the power to re-arrange physical particles in any way, and especially in such a way as to cause them to reflect light overawingly? It is incontrovertible that a very peculiar, commanding light is brought into the face by the activity of the upper faculties in man. We are to explain this light and its effects, by studying man as an organic multiplex. The light is there, and you know it is there. We see it. It is a physical fact.

On the other hand, there is an earthy, opaque look. "O ye hapless two," says Carlyle of Charlotte Corday and Jean Paul Marat, "mutually extinctive, the Beautiful and the Squalid, sleep ye well in the Mother's bosom that bore you both. This was the History of Charlotte Corday; most definite, most complete; angelic, demonic; like a star." (*French Revolution*, vol. ii., book vi., chap. i.) Compare the faces of Charlotte Corday and Marat.

Certain passions give a dark look to the countenance. How do they do that? Is it merely by a re-arrangement of the ultimate atoms of the skin and of the external parts of the eye? The astute materialist admits that certain emotions are accompanied by such displacements of the atoms of which the body is composed as permit the exterior of the countenance to reflect light only imperfectly. How is it that the bad passions thus relax us? It is incontrovertible that earthy passions give an earthy look to the countenance. The bestial man acquires an opaque and peculiarly repulsive complexion.

When I stood once in the Jewish Wailing-place in Jerusalem, and contrasted the pure blood of the Jew with the coarse blood of the Arab, I had before me on the one hand, countenances singularly capable of illumination; and on the other, faces singularly incapable of it. Say, if you please, that I am going off scientific ground here. I affirm that I have a scientific right to take the monogamistic Jew and the polygamistic Arab, or the Old Testament and the Koran put into flesh and blood by long centuries of experience, and to compare them. Not a few children from some of the best Jewish families on the earth are sent to Jerusalem for education; and even the careless observer of the faces of many of them must see that they are pure in blood, and, as I was compelled to think, of finer grain than the Italians and the Greeks of the Forum and the Acropolis. But, I said, "You have forgotten the English; you have forgotten the Americans;" and, as my conclusions were taking that posture, there came into the brown, crowded square two children in English dress, and began to converse with the Jewish children. I thought, "These are sons of rough men probably: they do not represent the English or the American fineness." They were superior in animal force, but plainly inferior in capacity for the solar look, to the Jewish boys with whom they conversed face to face. I asked to whom the outrivalled children belonged, and found they were sons of one of the most cultured men, indeed, of one of the missionaries, in the Holy City. The Arab, however, was the greater contrast, — opaque, repulsive, conspicuously impervious to light in his countenance; while, in the best specimens, the Jew shone from behind his physical integument at times like a light behind thin translucent marble. We know that this contrast exists in different men we meet, and in different moods of the same individual. Men may be made of floss-silk, and have æsthetic luminousness in their faces, and yet no solar light. The darkness of the Ethiop face does not hinder it from exhibiting either the solar light or its opposite. It is a wholly incontrovertible fact, that an earthy look comes from an earthy mood, and a solar look from a conscientious.

But now, will any one who reveres the scientific method deny my chief proposition, that the earthy look, other things being equal, quails before the solar? Is not that known to ordinary observation? No doubt, if a Cæsar or a Napoleon comes before some man of weak will, the latter, although he may be a good man, — and especially if he is a goody, a very different thing, — will quail. But give the latter the executive power and intellect of your Cæsar, and what is the result? *Other things being equal, Cæsar's eye goes down whenever it meets and does not possess the solar look.* The veriest sick girl with this solar light behind her eyeballs is more than a match for Cæsar without it. Yes, Cromwell's daughter was a match for him once, and Cæsar's wife for the man whose finger-tap overawed a Roman senate. There are no forces known to the lights of the eyes, that, other things being equal, ever do or can put down the solar light, even in the sick and the weak. Poets have celebrated many lesser radiances, and occasionally this highest radiance that can belong to woman. There are behind it an awe, and a right to command, which distinguish it from all other lights.

We know that the brute sees the sunset; but does it feel its pensiveness? No doubt the monsters that tore each other in the early geological ages beheld the risings of the suns, and their noons, and their descendings. The eyes of many a winged creature in the night reflect as perfect images of the stars as did Newton's. But do they appreciate what we call beauty, or sublimity, or natural law? *The world is a sealed book to the brute; and an archangel would say that it is to us.* On his vision, were he in the world, might fall no more than on ours; but he would read as many more meanings than we, as we than the brute. What is the significance of this mysterious, commanding, solar light? It is a visible fact, but we gaze on it apparently with brutish, uncomprehending eyes. We do not intellectually fathom it, and yet we feel it much as the brute feels the authority of the human eye.

Your pensive, wailing, inferior creature gazing into the human face seems very often to be governed by an awe that does not arise from fear of physical injury. There is command in the intellectual light when it is contrasted with the merely animal light. The poor four-footed brute goes away with, it may be, a vague sense of worship, or of affection at last, if you draw it towards you. The canine creatures can thus be tamed; and untamable beasts can be looked out of countenance,—even your lion and your tiger, if you gaze steadily upon them, contrasting the human radiance with the animal. Now, just as that four-footed brute may feel, looking into your eyes, so I confess I have felt sometimes when looking into the eyes of those better than myself. I have felt brutish; I have felt my inferiority; I have quailed, I confess it, before eyes which I thought had behind them a holier light than mine have ever shown. I sometimes compare my mood at such instants with that of your creature that cannot speak, and that slinks away with a sense of inferiority. I know that this light is my master. I do not quite understand the light. The poor brute does not understand the radiance in the human eyes, but confesses that this light is above it; and so I have felt that in the solar radiance there is something above all my earthiness. There is no man that can look on what we call the solar light in the human countenance, and feel that it is genuine, and not reverence it. There is a natural awe in its presence. What does this incontrovertible fact mean? There are only a few animals so low that they cannot be looked out of countenance; and there are only a few men so low that they cannot be looked out of countenance also.

As the brute sees the sunset, and does not understand it, gazes upon the glory and beauty, and finds it a sealed book, so we see but do not appreciate this marvellous capacity of man's countenance to clothe itself in solar light; and yet in it we are looking upon something which in another age will be better understood in the name of science. So much is already incontestably known: that the solar light exists; that all other light quails before it; that it springs from the heights of conscience; and that the only complete and the only victorious self-culture must be solar self-culture. Even if we were compelled to pause here, we should have attained a point of vision where, as Goethe said when he climbed Vesuvius, one look backward takes away all the fatigue of the ascent, and is a regenerating bath. Our age believes in culture; a more scientific age will believe in solar self-culture. On the height to which our inductive research has now carried us will be erected tabernacles to the honor of the only culture by which, under natural law, the yet opaque face of civilization can find transfiguring and commanding radiance.

What of the Transfiguration? Was that an example of solar light? The clouds are slowly parting above this theme, and is it possible that we have not yet reached its summit? Is this outlook of ours only from a mountain range so low as to be hardly a vestibule? There is a solar light; and what if, adhering now to all that science proves concerning it, we gaze up the Alps, so unexpectedly uncovered, as the vapors part themselves above the stupendous veiled summits of revelation? Is it possible that their height itself has kept them obscured until we had little knowledge of their existence?

I am asking you here and now only to take scriptural facts as statements of the Christian point of view. If there is any man here who regards the history as mythical, even he will allow me to use it to show what Christianity believes. I am scientifically authorized to make reference to it all to indicate what has been taught on the topic of the solar radiance.

It is recorded that in an Eastern city a martyr was once tried, and as all they who sat in the council looked steadfastly on him they beheld his face, as it were the face of an angel. Is it possible that the solar light present in this case, and in approximately similar cases in our day, is the same thing in each? It is recorded also, as we remember, now that we allow our minds to sweep through the vistas of historical examples, that a lawgiver, who yet rules the centuries, once had, as he came down from a certain mount, a face that shone. The old Greek used to inquire with intensest philosophical interest what that light was which appeared once, not in the face only, but in the hands and in the feet and in the garments, of the only Member of the human race who has ever shown us solar light at its best. The Greek asked in the early days of Christianity, Whence that light? There is incontrovertibly a solar light which fills the faces of a few men and women in our day. Dante, I take it, is looking towards this fact when he says, "That which in heaven is flame, on earth is smoke." Is it possible that the solar look which comes into the countenance whenever the loftier zones of feeling are in full action is of the same sort with that which appeared in the face of Dante's Beatrice, delighted to do good; and in the face of Him whose countenance was like that of an angel; and in the face of Moses; and in the unfathomed symbolisms of the Transfiguration? Is it of the same sort with that light which fills the world of those who have no need of the sun, because the face of the Lamb doth lighten them, and the glory of God is the lamp of their tabernacle?

These questions may well blanch the cheeks; but they are to be studied in the spirit of science, if we are to think with any freedom or breadth. Surely, here is a train of investigation not often followed in detail. Having read to you twelve propositions drawn up from the point of view of science, let me read twelve drawn up from the point of view of unadulterated Christianity.

1. It is historically known that the early Christians regarded the possession of the solar, commanding look, as a sign of the possession of the Holy Spirit. Stephen, when full of the Holy Ghost, had a face like that of an angel. When Moses came down from the mount his face shone.

2. At the Transfiguration this solar light had its supreme manifestation.

3. That light was perhaps a revelation of the capacities of the ethereal enswathement of the soul, and of a spiritual force acting through the physical organization.

To those who were present at the Transfiguration, the Cross did not seem other than the voluntary humiliation of Him who was stretched upon it. A revelation of some of the capacities of the spiritual body, which death, according to Ulrici, separates from the flesh, was made to three of the disciples; and they were the three who afterwards witnessed the agony in the garden, and were nearest to the Crucifixion. They were prepared for the witnessing of the agony by the previous revelation of the glory of the body which was transfigured.

There was a cloud which appeared in the Transfiguration, and it is recorded that the disciples feared as they entered into that cloud. It is said, also, that when He who was transfigured walked once up the slope from Jordan to Bethlehem, the disciples followed him, and were afraid. The light which appeared in the Transfiguration appeared again in the Ascension; and the cloud that overshadowed the former was the chariot of the latter.

We have considered here (*Biology*, Lecture XIII.) the schemes of thought which assert that there may be three things in the universe, and not merely two: matter and mind, and a middle somewhat, ordinarily called the ether, and at least not atomic, as what we call matter is. We know how Ulrici and others speak of an ethereal enswathement of the soul, and of a spiritual body.

What if the cloud which appeared at the Transfiguration was some revelation to the human sense of that ether which Richter calls the home of souls? What if the transfiguring light was but a revelation of the capacities of the spiritual being, enswathed within the flesh as light is enswathed within the fleecy tabernacle of the translucent flying clouds in the noon yonder above our heads? Mysterious, you say? But after all we must adhere to the principle that every change must have an adequate cause. As Dante says, there is smoke on earth; the solar light in the human body is dim here; but what is this flame, when at its best? The light of the fire that shines in the eyes of a good man or woman, how bright would it be if their goodness could be enlarged to the measure of that of the Soul that never sinned? How would it illuminate then the whole frame? Is there unity of kind between the light that we call the solar look in scientific parlance, and the radiance that filled Stephen's face, or that of Moses? A spiritual force was concerned in the two cases, and its powers are yet unchanged. Was not the same force concerned in the Transfiguration also? Was not one object of that event to make a revelation of the hidden glory of our Lord's Person?

Are we going too far when we say that these topics which interested the old Greeks so passionately are worth looking at as the vestibule to the majestic temple of conscience? Activity of the upper zones of feeling is what causes this peculiar light in our little experience of it. We have but the twilight, a dim scintillation of this radiance. But we know that what little we have of it comes from the innermost holiest of conscience. Raphael studied the Transfiguration, and his painted conception of it was borne aloft above his funeral bier. Are we not in the advances of science obtaining some views of it which his canvas cannot show us?

It is recorded of our Lord, that, as he prayed, the fashion of his countenance was altered, and his face did shine as the sun, and his raiment became white and glistering, so as no fuller on earth can white them.

4. As our Lord's body was human, it is not too much to say that its mysterious, overawing capability of receiving illumination from within by spiritual forces must be supposed to be possessed in some degree by every human body.

5. An obscurest form of perhaps the same solar light is yet seen occasionally among men.

6. We know that the light arises from the blissful supremacy of conscience, and the activity of all the higher powers of the soul.

7. As the Scriptures made the possession of this light one of the signs of the possession of the Holy Spirit in the scriptural days, we must infer that this light is such a sign in these days.

8. The innermost holiest of conscience, in blissful supremacy, is therefore known to science as well as revelation, as the temple of the Holy Spirit.

9. But the Holy Spirit was shed forth by Him who was the Light that lighteth every man that cometh into the world.

10. The modern solar light and that Light are therefore identified.

11. But the solar light is scientifically known to be the only commanding light; and therefore the Light that lighteth every man that cometh into the world is scientifically known to be the only commanding light.

12. The only complete and the only victorious self-culture is scientifically known to be solar self-culture; but solar self-culture and Christian self-culture, so far forth as both are solar, are identical; and both are known to science as solar so far forth only as they originate in the innermost holiest of conscience.

Harvard yonder, Matthew Arnold, Stuart Mill, all ranks of modern scholars, believe in culture. But there is only one form of culture that gives supremacy, and that is the form which produces the solar look; and the solar look comes only from the Light that lighteth every man that cometh into the world. It may be incontrovertibly proved by the coolest induction from fixed natural law, that the highest culture must be that through which the solar look shines, and that this look is possible only when there exists in the soul glad self-surrender to the innermost holiest of Conscience. In that innermost holiest, Christianity finds a personal Omnipresence. Culture should believe in the law of the survival of the fittest. Two lights conflict, — the earthy and the solar. Your eyes filled with poetic rapture, your loftiest attitudes of merely æsthetic or intellectual culture, quail, other things being equal, before the solar look. Here is a fact of science: a visible, physical, haughty circumstance of yet unfathomed significance; an unexplored remainder on which what calls itself culture, and quails, may do well to fasten prolonged attention.

> "Satan . . . dilated stood,
> Like Teneriffe or Atlas unremoved;
> His stature reached the sky, and on his crest
> Sat Horror plumed. . . .
> The Eternal . . .
> Hung forth in heaven His golden scales, yet seen
> Betwixt Astræa and the Scorpion sign.
> . . . The fiend looked up, and knew
> His mounted scale aloft; nor more; but fled
> Murmuring, and with him fled the shades of night."
>
> MILTON, *Paradise Lost*, iv. 985.

[Applause.]

III.

THE PHYSICAL TANGIBLENESS OF THE MORAL LAW.

THE EIGHTY-THIRD LECTURE IN THE BOSTON MONDAY LECTURESHIP, DELIVERED IN TREMONT TEMPLE, OCT. 15.

Quid enim aliud est natura, quam Deus, et divina ratio, toti mundo et partibus ejus inserta? — SENECA: *De Benef.*, iv. 7.

So lange das Wort Gott in einer Sprache noch dauert und tönt, so richtet es das Menschenauge nach oben auf. — RICHTER: *Levana.*

III.

THE PHYSICAL TANGIBLENESS OF THE MORAL LAW.

PRELUDE ON CURRENT EVENTS.

THE parliamentary expenses of the Brighton railway in England were fifteen thousand dollars a mile. George III. sometimes expended for purposes of political corruption the money voted to him as king, and called his gifts golden pills. We all remember very well that Lord Chatham's measures of reform were often spoiled by Lord Bute, and that the latter frequently succeeded by striking the great statesman's followers with a golden club. It is said that Lord Bute, in a single day, issued to the order of his agents twenty-five thousand pounds. On one occasion a government loan was raised among his adherents by private subscription, on such terms as to distribute among them three hundred and fifty thousand pounds of public money. In the days of the Pensioned Parliament, peerages were bought and sold, and now and then the amounts paid for them entered in the books of the exchequer. It was very common to buy a member of the Lower

House, and even a lord was sometimes sold over his chair as you sell goods over the counter of a stall. It is altogether too early yet to forget political corruption in England; but since the reform measures of 1832, civil-service amelioration has taken such hold of Great Britain, that it is now almost an unheard-of procedure to sell, or to attempt to buy, a member of Parliament. The corruption which existed in Great Britain during the railway mania was perhaps as great as that in the United States in the times of our Credit Mobilier. During the struggles with Napoleon, corruption in English public life was far-reaching in every political department. Macaulay says, however, that even then the judiciary was not corrupt in England, and that commerce was generally very sound. It is to be remembered that we have an elective judiciary in twenty-two States, and that probably our miserable civil service has affected the judiciary more than the judges were ever influenced in England by political corruption. Nevertheless there was a nobility in England, depending largely on the civil service for places for sons not put into position by the law of primogeniture. Second sons, third and fourth, and so on, were to be pensioned in a state church or in a political office, or in the army or navy. If our judiciary is a more corrupt body than the English ever was, we have no upper class with strong interests at stake in the existence of corruption. Therefore the field is perhaps not a very much more difficult one here now than it was in England in 1832, for the progress of civil-service reform.

How, then, did this change occur in England? A Congress meets to-day at Washington, called together by the first American President who has attacked what George William Curtis calls the "consuming gangrene" of our public life, office-holding control of politics. [Applause.] This English history, this black page and the present white page, are they not worth attention from Congress and from us? Did the black page immediately precede the white? or were there some gray leaves interspersed, some blotched and almost ragged pages, between the dark record of corruption and the present honest civil service in England? As early as 1832 reform began; but it was not until about the year 1853, when Sir Stafford Northcote drew up his definite propositions, that civil service reform grew to be a victorious cause in England. There have been, however, twenty years of crescent success in Great Britain for civil service reform. The result is, that to-day the contrast of American and English politics is vastly to our disadvantage, while the contrast of American politics under Washington and Jefferson with English politics of the same period would have been greatly to the disadvantage of the English. About the time when the reform measures were passed in Great Britain, Andrew Jackson introduced here the spoils system. And now that twenty years of vigorous action on the part of the executive of Great Britain has shown what can be done for civil service reform there, why should we not cast a sharp glance upon that page of English precedent, when the topic of civil-service reform comes before America, with its fatter and vaster political spoils, as a question almost of life or death?

What is the particular regulation of office-holding in Great Britain? The premier appoints, of course, his colleagues in his cabinet, with the advice of the King or Queen. Then the cabinet together choose subsidiary officers just under them. Only about thirty men in the upper ranges of the civil service are changed when the party or the ministry changes. With very few and now decreasing exceptions, the lower ranges are filled by competitive examinations. A man once in position expects to keep his place during good behavior, and to be promoted for merit. The consequence is, that the control of politics has been taken out of the hands of party in Great Britain, so far as office-holding is concerned, and put into the hands of the people, where it belongs. To-day public sentiment probably has a greater power over parliamentary action than over congressional: at least, its effects are more immediately perceived. A change can be brought about more quickly in the Parliament than in Congress by a haughty, commanding public sentiment. The reason is that patronage is not left in the hands of members of Parliament to corrupt the country through every small office.

We must beware of demagogues who clamor against an office-holding aristocracy, and who assure us that civil-service competitive examinations would result in the institution of a class having peculiar privileges. That class in England serves Lord Beaconsfield to-day and Gladstone to-morrow. How peculiar are its privileges? on which side is it? It is a great profession; it has learned how to do its work; it keeps in place although ministries change. Just so, if we had such an office-holding class in this country, it would serve both political parties, do its work well, and could not be bought and sold from custom-house to post-office, or become a standing bribe in Congress. We must not allow the office of an American senator to become a gift enterprise. As a reformed civil service would be filled by merit, and as competition for places in it would be open to everybody, we should have a class serving both political parties, and therefore no aristocracy at all.

We ought to conduct the mechanical part of our governmental work as a great factory does its business, by retaining the servants who have shown themselves capable. When ten or twelve acres of factory floors change owners, the shrewd men in Boston and New York who manage the enterprises that move the whirring looms on those floors do not change all their foremen nor all their operatives. They know what men have done well, and keep them in place. Our national business is to be managed for the benefit of the nation, and not for that of a party. [Applause.] It is to be managed by the people that own the whirring looms, and not by the men who are speculators at the best, and who make a business of fleecing each other as rivals. Of course there will never come, in America, any peace or purity in politics until the day of the disestablishment of the machine in politics. [Applause.]

THE LECTURE.

After Robespierre had choked the Seine with the vainly whimpering heads sheared away by the guillotine, there came an hour when a death-tumbril containing himself was trundled toward the fatal French axe. Carlyle tells us that the streets were crowded from the Palais de Justice to the Place de la Revolution, the very roofs and ridge-tiles budding forth human curiosity, in strange gladness. The soldiers with their sabres pointed out Robespierre, as the crowd pressed close about the cart. A French mother, remembering what rivers of blood that man's right hand had wrung out of the throat of France, springs on the tumbril, clutching the side of it with one hand, and, waving the other sibyl-like, exclaims, "Your death intoxicates me with joy." The almost glazed eyes of the would-be suicide Robespierre open. "*Scélérat*, go down, go down to hell with the curses of all wives and mothers!" A little while after Samson did his work, and a shout raised itself as the head was lifted, — a shout, says history, which prolongs itself yet through Europe, and down to our day. (CARLYLE, *The French Revolution*, vol. ii., book viii., chap. vii., "Go down to.") That word "*down*" will never be understood by us until we contrast it with the "up," with which men salute the Gracchi and the Phocions, the Lafayettes, the Washingtons and the Hampdens, and which prolongs itself mysteriously in history. The word "down," once uttered by the ages, is rarely reversed; and the word "up," once looking haughtily on that word "down," very rarely, in history, changes its countenance.

There appear to be behind these two words inexorable natural laws. Is it possible to discover any of them?

1. Instinctive physical gestures accompany the action of strong feelings.

2. It is a peculiarity of the strongest moral emotions, that the general direction of the physical gestures which they prompt is either up or down.

3. By the operation of a fixed natural law of the human organism, we hang the head in shame or acute self-disapproval.

4. By the operation of a fixed natural law, we hold the head erect when conscious of good intentions, or acute self-approval.

5. It is a physical fact, demonstrable by the widest induction, that the gestures prompted by the blissful supremacy of conscience have their general direction upward, and give the human form a reposeful and commanding attitude.

6. It is also a physical fact, demonstrable by the widest induction, that the gestures prompted by the opposite relations to conscience have their general direction downwards, and give the human form an unreposeful and more or less grovelling attitude.

7. Other things being equal, the latter attitude alway quails before the former.

8. By fixed natural law the upward gestures induced by an approving conscience and the activity

of the higher faculties are accompanied by a sense of repose, of unfettered elasticity, and of a tendency to physical levitation.

9. By fixed natural law the downward gestures induced by a disapproving conscience are accompanied by a sense of unrest, of fettered activity, and of a tendency to delevitation.

10. In some of the most celebrated works of great artists, the human form is represented as in a state of physical levitation; but this is always pictured as accompanied and caused by the blissful supremacy of conscience and of the higher faculties.

11. It will be found, on an examination of personal consciousness, that there is in the artistic sense a feeling that forms exhibiting the blissful supremacy of conscience and of the higher faculties will float, and that forms which do not exhibit these traits will not.

12. So deep is the instinct concerned in the upward gestures produced by an approving, and the downward produced by a disapproving conscience, that history contains large numbers of alleged instances of the physical levitation of the human form in moral trance.

13. Without deciding whether these cases are authentic facts or not, their existence shows the intensity of this instinct, and the great significance of the inexorable natural law which it reveals.

14. In the existence of the instinctive upward and downward physical gestures accompanying the approval and disapproval of conscience, natural law reveals a distinction between up and down, higher

and lower, in moral emotion; and, in doing that, founds an aristocracy, strictly so called, or government by the best, and determines that it shall rule; and these instinctive gestures, occurring according to natural law, are a proclamation of that aristocracy, — the only one recognized by nature, and the only one that will endure. [Applause.]

15. It will be found that all the instances in human experience of the distinction between up and down and higher and lower, as thus defined by observation, may be summarized under a law of moral gravitation proceeding from conscience.

16. Moral gravitation, therefore, is as well known to exist, and is as tangible, as physical gravitation.

17. But all law in nature is but the uniform action of an Omnipresent Personal Will.

18. The tangibleness of the moral law in conscience is scientifically known, therefore, to be identical with the tangibleness of an Omnipresent Personal Will.

19. Moral gravitation is thus *in*, but not *of*, the soul.

20. There is, therefore, in man a Somewhat or Some One not of him, and spiritually, and in a significant sense physically, tangible through conscience.

Ascending that stairway of propositions, I have not asked you to pause to converse on the balustrades; but, assuming that we have gone up the height together, let us, now that we stand here, look back, and make sure that all our steps were on the adamant. Take no partisan witness, however, in our examination of this case before the learned jurors in this assembly. You say that I am a lawyer making a plea for a foregone conclusion! Is William Shakspeare a partisan? Did he know any thing of human nature? The *heaviness* of the soul of a man who has done evil, is that recognized by William Shakspeare?

Imagine that this Temple is Bosworth battle-field. There is the tent of Richmond, and here the tent of Richard. William Shakspeare shall guide us in our study of natural laws in these two tents. He does not look through partisan lenses. He is no theologian. What are these forms which rise in the dead midnight between the two tents? There are eleven ghosts here. Shakspeare is behind every one of them. They utter nothing that he does not put into their lips; when they speak, he speaks; and some of us have been taught to believe that when Shakspeare speaks, Nature speaks.

> "Let me sit *heavy* on thy soul to-morrow;
> Think how thou stabb'dst me in my prime of youth,
> At Tewksbury: despair therefore, and die."

So speaks the first ghost at Richard's tent.

> "Be cheerful, Richmond; for the wrongèd souls
> Of butchered princes fight in thy behalf:
> King Henry's issue, Richmond, comforts thee."

So speaks the same ghost at Richmond's tent.

> "When I was mortal, my anointed body
> By thee was punchèd full of deadly holes.
> Think on the Tower and me; despair and die:
> Harry the Sixth bids thee despair and die."

So speaks the second ghost at Richard's tent.

"Virtuous and holy, be thou conqueror!"

So speaks the same ghost at Richmond's tent.

"Let me sit *heavy* on thy soul to-morrow, —
I that was washed to death with fulsome wine,
Poor Clarence, by thy guile betrayed to death!
To-morrow in the battle think on me,
And fall thy edgeless sword; despair and die."

So speaks the third ghost at Richard's tent.

"Good angels guard thy battle! Live and flourish!"

So speaks the same ghost at Richmond's tent.

"Let us sit *heavy* on thy soul to-morrow."

So speak the ghosts of Rivers, Grey, and Vaughan, at Richard's tent.

"Awake, and think our wrongs in Richard's bosom
Will conquer him! — Awake, and win the day."

So speak the same ghosts at Richmond's tent.

The ghost of Hastings rises. The ghosts of the two young princes rise.

"Dream on thy cousins smother'd in the Tower.
Let us be *lead* within thy bosom, Richard,
And *weigh thee down* to ruin, shame, and death!
Thy nephews' souls bid thee despair and die. —
 Sleep, Richmond, sleep in peace, and wake in joy:
Edward's unhappy sons do bid thee flourish."

The ghost of Queen Anne rises.

"Richard, thy wife, that wretched Anne thy wife,
That never slept a quiet hour with thee,
Now fills thy sleep with perturbations.
To-morrow, in the battle, think on me,
And fall thy powerless arm; despair and die. —

Thou, quiet soul, sleep thou a quiet sleep,
Dream of success and happy victory;
Thy adversary's wife doth pray for thee."

The ghost of Buckingham rises.

" The first was I that helped thee to the crown:
Oh, in the battle think on Buckingham,
And die in terror of thy guiltiness!
God and good angels fight on Richmond's side;
But Richard fall in height of all his pride."

The ghosts vanish.

Is this natural? or supernatural? or both, and the one because it is the other? [Applause.]

Your Richard wakes yonder in his tent.

" O coward conscience, how thou dost afflict me! —
The lights burn blue. It is now dead midnight:
Cold fearful drops stand on my trembling flesh.
I am a villain; yet I lie, I am not.
Fool, of thyself speak well; fool, do not flatter.
My conscience hath a thousand several tongues,
And every tongue brings in a several tale,
And every tale condemns me for a villain.
Perjury, perjury, in the high'st degree;
Murder, stern murder, in the dir'st degree;
All several sins, all used in each degree,
Throng to the bar, crying all, Guilty! guilty!
I shall despair. — There is no creature loves me;
And, if I die, no soul will pity me; —
Nay, wherefore should they, — since that I myself
Find in myself no pity to myself?
Methought the souls of all that I had murdered
Came to my tent; and every one did threat
To-morrow's vengeance on the head of Richard."

King Richard III., act v., sc. iii.

PHYSICAL TANGIBLENESS OF THE MORAL LAW. 73

Let me sit *heavy* on thy soul to-morrow! So spoke Shakspeare; so, the ghosts; so, inductive science; so, natural law; so, that Somewhat which is behind all natural law; and so, that Some One who is behind the Somewhat. [Applause.]

You will allow me to make reference here to some of the subtlest of unexplored human experiences. I am by no means drifting out of the range of scientific currents and received thought, even if I venture to sail boldly into the fog which lies along the shore of many an undiscovered land. But, my friends, put Shakspeare at the helm. Let us recognize him as the pilot; and, remembering what weight he puts upon the word *heavy*, dare to look into the canvas of a Raphael and an Angelo a moment, and into this deeper canvas of our own souls, painted by natural law, that is, by the fingers of the Personal Omnipresence, who was, and is, and is to come. I affirm, what no man can deny, that the natural language of gesture is God's language. We did not invent it. Surely natural language is the language of nature; and these gestures which make us hang the head, and give us the erect attitude, are proclamations made, not by the will of man, but by the will of that Power which has co-ordinated all things, and given them harmony with each other, and never causes an instinct to utter a lie. We have heretofore looked carefully into the distinction between an organic and an educated tendency. It would mean very little if men had been taught to hang their heads in shame. It would mean very little if men by a process of education had learned to assume the erect attitude when conscience is supreme. It is scientifically sure, however, that, when an organic instinct can be discovered, we have a right to infer from its existence that of its correlate. We know that where there is a fin, there is water to match it; where there is a wing, there is air to match it; an eye, luminousness to match it; an ear, sound to match it. The migrating swans fly through the midnights and the morns, and they lean in perfect confidence upon the Maker of their instinct, knowing that if God has given them a tendency to fly to the South, he will have provided a South as a correlate to the tendency. (Our great tests of truth are: intuition, instinct, experiment, and syllogism.) Incontrovertibly we have organic, and not merely educated tendencies concerned in these instinctive gestures, by which conscience in blissful supremacy gives the human form a commanding or overawing attitude, and sometimes a levitated mood. I say that the mood is levitated, whether the form is or not.

In certain highest moments, when conscience assures us that the stars fight for us, we do have a feeling that if cast out unsupported into the ether we should float there; and we have at other times a feeling that if we were disembodied, and cast out into the unknown, we should sink. These two subtle and subtly contrasted organic feelings are endlessly significant.* Do you believe the forger, the perjurer, the murderer, has any feeling that he could float aloft with the great levitated forms which the artists have put upon canvas? After studying often at Dresden Raphael's Sistine Madonna, who will float, I paused in the Louvre many times with dissatisfaction before Murillo's Madonna, who will not. She stands on a crescent moon, and I think she needs it as a support. But the Venus di Milo will float, although she is in marble. We have these instinctive feelings, although we do not understand them any more than the brute does the sunset. We cannot rid ourselves of them if we allow our thoughts and emotions to follow a natural course. We have a strange, deep sense by which we authorize ourselves to say of now and then a female form in art, and even of the male form occasionally, though oftener of the female, that it would float if left alone in the ether. This instinct is an indisputable fact. It is surely a shore, although veiled yet in vapor. We have not approached that coast much yet, but there is the instinct; there is firm land here, and the trend of its beaches, where lies so much undiscovered gold, must be in perfect accordance with that of all these instinctive gestures. Begin with what cannot be controverted, or the proposition that we hang the head in shame, and hold it erect in conscious self-approval. We know that some attitudes, in deep remorse, bring a man down almost to the posture of the brute. We grovel in the dust at times, when we feel ourselves under the full thunder and lightning of the moral law. Mr. Emerson says that he has read in Swedenborg — he means he has read in natural law — that the good angels and the bad angels always stand feet to feet; the former perpendicularly up, the latter perpendicularly down. If you please, that is science: it is not poetry. It is poetry; but it is science, too. We see a gleaming curve of the law in the hanging head and in the erect and reposeful and commanding attitude. We see it in that sense of elasticity and almost of physical levitation which arises in states of moral trance. We see it on the canvas of great painters in yet higher manifestations; and when we come to the asserted cases of physical levitation, we have at least an indication of the intensity of the instinct they represent, and therefore of its value as a scientific guide.

Shakspeare is at the helm. Walk forward into this wheeling vapor, and gaze shoreward from the bow of the vessel. Let him keep his place. He will not ground you upon any rocks or shoals. Go to the vexed leeward rail nearest this strange shore sounding there under this obscuring mist, and open as a chart — what? Why, the British Quarterly Journal of Science, edited by Professor Crookes. What does he say? Has he any guide-book to this fascinating unknown coast? He publishes careful articles, in which are summed up a large number of the alleged historical cases of levitation in moral trance. Pliny in his Natural History (vii, 18) said long ago that the bodies of all living things weigh less when alive and awake than sleeping or dead. (Mares præstare pondere; et defuncta viventibus corpora omnium animantium, et dormientia vigilantibus.) Dean Trench

(*Notes on the Miracles*, ed. vii. p. 289) defines man as "the animal that weighs less when alive and awake than dead or asleep." It is well known that the levitation of the body of Mr. Home in London is asserted on the testimony of eye-witnesses, including in their number Professor Crookes, editor of the Quarterly Journal of Science, Lord Lyndhurst, and many other men of large experience, trained minds, full culture, and unimpeached integrity. On a single page of the guide-book to which I have referred you (Quarterly Journal of Science, January, 1875, p. 53), you will find a statement of the names, country, condition, and date of life, of forty <u>levitated persons</u>. "The darker and less historical the age," says this writer (p. 52), "the more miracles, but the fewer of these phenomena [of levitation]. The testimonies to these, absent so far as we can see in the ages from the fourth century to the ninth, increase in number, respectability, and accuracy, from the latter to the present day." In this long list of instances, the levitations occur as a rule in states of moral elevation, or trance. "If levitation has occurred," says this authority, "it is natural. Under what conditions, we may never be able the least to define; but whatever happens we must call natural, whether the naturalness be clear to few or many, to none or all of us (p. 39)." Professor Crookes thinks that if we can prove that Cæsar was assassinated, we can prove that there have been cases of levitation. I do not agree with him. I think it very doubtful whether we can now demonstrate that physical levitation has occurred under the eyes of experts, or can be proved to the satisfaction of men of science. But this fully accredited teacher has a right to be heard in the majestic roar of the unconquered surf of this unknown coast. Shakspeare is there at the helm; he will draw the ship off in a moment; but you must peer once in the name of science, and of more than one advanced pilot of modern thought, into this mist. Professor Crookes affirms that if we are to be candid students of history, we shall be very shy of denying that there never has been physical levitation, as it is sometimes represented on the canvas of our great painters. Personally he has no doubt that it occurs in states of moral trance.

We know something of what it is to be elastic when we feel that we are right with God and man; and that fact is a deep glimpse into this wheeling, smiting mist. It is surely worth while, gazing in the direction of this gleam of analogy and fact, to ask whether there have been cases in which the human form, under the highest activity of conscience, has been lifted aloft. I do not ask you to accept Mr. Crookes's statements. I ask you only to note what some leaders of the very latest science are saying, and to keep an eye on the lee shore, meanwhile taking soundings every now and then. Keep well away from the rocks of Spiritualism. [Applause.] There are Mahlstroms in which, listening, it may be, to evil spirits, man sometimes mistakes the moral downward for the moral upward; and, gazing into the azure of the wide, swift, smooth, circling sea at the whirlpool's edge until dizzy, persuades himself that its inverted reflection is the sky; wishing two wives, takes some gleam of a lie out of that lower azure as his justification for having them; adopts the Mahlstrom, in all its downward swirls, as an upper heaven, and so plunges into its glassy throat, as if he were ascending. Keep out of that. [Applause.]

Nevertheless I cannot discuss the topic of uncontroverted physical facts concerning conscience, without asking you to notice in the name of Shakspeare and all the common instincts on the one hand, and of all of the latest research on the other, that a physical tendency to levitation is a matter worth investigation.

But now, my friends, even if we could not make any use of Mr. Crookes's facts, we do know how tangible the moral law is. We know that these gestures upward and downward reveal subtle arrangements in the connection of our organization with conscience; that they indicate instincts; and that all instincts have their correlates. Suppose that I could take you no farther up this staircase, along its twenty steps, than to the tenth or fifteenth. Suppose that we cannot go up together over more than half these steps: you who stand on the lower platform will yet, when you look back, have an outlook worthy of study. I know that I have an instinct by which my gestures, in the midst of conscientious self-approval, express command, repose, elasticity; and that when conscience is against me I grovel naturally. Up and down are words physically proclaimed by natural law. There is no reversing the relations of the peerage of heaven. I want the culture that will bring me near to the Court. I therefore must studiously examine the only steps by which man can ascend toward the gates that have foundations. I know that selfish pride and self-approval through conscience are as different as east and west. They are so far apart that east and west, compared with them, have nearness and cohesion. A reposeful mood and peace are given by a blissful supremacy of conscience, but these are rarely conscious of themselves, as pride always is. If the face has a solar light, it is usually unconscious of the possession of that radiance. And so, if a man have the approval of conscience, if the upper nature be in blissful supremacy, he is usually unconscious of his mood. No emotion has its full strength until it is so profound that its possession is not noticed by its owner. We are not fully given up to any feeling until we not only have possession of it, but become unconscious of the sorcery by which it possesses us. The orator must not only have possession of his subject, but his subject of him. When it has possession of him, you are not conscious of him, nor is he of himself, but only of his theme.

If I were able to go up only half the steps that you have ascended here with me, I should feel myself other than an orphan in the universe. We ask how God can be touched. How can we come near to the ineffable Somewhat and Some One, that lies behind natural law? We are poor flowers opening toward the noon. We have no eyes to see, and yet we have nerves to feel. Do we need any thing more? We are sure that we have the nerves, and that we touch the sunlight. We know scientifically that there are an up and a down in natural law in its moral range. We are as conscious of this moral gravitation as we are of physical gravitation. We touch a Somewhat that lifts us, and the absence of which leaves us to sink to what appears to be a pit bottomless; and we know that this gravitation is a natural law. But it is a truth of science, that every natural law is the constant operation of an Omnipresent Personal Will; and, therefore, in the incontrovertible physical facts illustrating moral gravitation as a natural law, have we not the touchings of the Personal Omnipresence, as much as the flower has the touchings of the sunlight when it absorbs its beams? [Applause.]

> As feel the flowers the sun in heaven,
>   But sun and sunlight never see;
> So feel I thee, O God, my God!
>   Thy dateless noontide hid from me.
>
> As touch the buds the blessed rain,
>   But rain and rainbow never see;
> So touch I Thee in bliss or pain,
>   Thy far vast Rainbow veiled from me.
>
> Orion, moon and sun and bow,
>   Amaze a sky unseen by me;
> God's wheeling heaven is there, I know,
>   Although its arch I cannot see.

> In low estate, I, as the flower,
>   Have nerves to feel, not eyes to see;
> The subtlest in the conscience is
>   Thyself and that which toucheth Thee.
>
> Forever it may be that I
>   More yet shall feel, and shall not see;
> Above my soul thy Wholeness roll,
>   Not visibly, but tangibly.
>
> But flaming heart to Rain and Ray
>   Turn I in meekest loyalty;
> I breathe and move and live in Thee,
>   And drink the Ray I cannot see.

[Applause.]

What of the Ascension? It is said, to turn now one glance upon the Scriptural record, that One, whose face did shine as the sun in solar light, and who illustrated that radiance as no other member of the human race has ever done since, as He blessed His disciples was lifted up from them, and a cloud received Him out of their sight. Will you quail here, when you see the perfect unity between the natural law, as I have endeavored to unfold it, and this action of spiritual forces in that Member of the human race, who, at the Transfiguration, illustrated the glorious capacities of the same forces to give to the present organic body solar light? I know that in *us* there is a levitating tendency in a moral trance. I know that as *we* pray, the fashion of our countenance is altered. And it is recorded that as *He* prayed, the fashion of His countenance was altered, and that as

He blessed his disciples He was borne up from them. Without controversy, great is the mystery of Godliness. God was manifest in the flesh, justified in the spirit, seen of angels, preached unto the Gentiles, believed on in the world, received up into glory. You say that I am treading here upon the very edge of blasphemy, in assuming that any natural law is concerned in these summits of revealed fact. But, my friends, the distinction between the natural and the supernatural is one that may be stated in many ways. The natural to me is merely God's usual action, the supernatural his unusual action. God's will is uniform; and if you and I experience some tendency to stand erect when we are right with God, if you and I have some tendency to spiritual levitation when we are in a moral trance, who shall say, if our goodness had equalled that of the Soul that never sinned, that we should not know what levitation is, as he did?

I am perfectly aware that I am venturing into unexplored remainders of thought, but it is my purpose to do so; for here, at the temple's opening in this structure which I am building, full of reverence for conscience, I wish to erect two pillars, — two gorgeous marble shafts, if you please to look on them as I do, facts of science making them glorious, — two columns, one on either side the door, — Solar Light and Moral Gravitation. Both are physical facts. Both we can touch in the lower flutings of the shafts; and we know by the argument of approach, and by the whole scheme of analogical reasoning, that if the solar light were carried up to its loftiest capacity, it might, at its summit, have the Transfiguration ; and if the laws of moral gravitation are examined, and we ascend them to the highest point to which analogy can take us up, we may, without violating, by the breadth of a hair, scientific accuracy, find there the Ascension. [Applause.]

IV.

MATTHEW ARNOLD'S VIEWS ON CONSCIENCE.

THE EIGHTY-FOURTH LECTURE IN THE BOSTON MONDAY LECTURESHIP, DELIVERED IN TREMONT TEMPLE, OCT. 22.

Und ein Gott ist, ein heiliger Wille lebt,
   Wie auch der menschliche wanke ;
Hoch über der Zeit und dem Raume webt
   Lebendig der höchste Gedanke,
Und ob Alles in ewigem Wechsel kreist,
Es beharret im Wechsel ein ruhiger Geist.
            SCHILLER: *Die Worte des Glaubens*, 4.

   Se Dio veder tu vuoi,
      Guardalo in ogni oggetto,
      Cercalo nel tuo petto,
      Lo troverai con te.

   E se, dov' ei dimora,
      Non intendesti ancora,
      Confondimi, se puoi ;
      Dimmi, dov' ei non è ?
            METASTASIO: *Betulia Liberata*, ii.

IV.

MATTHEW ARNOLD'S VIEWS ON CONSCIENCE.

PRELUDE ON CURRENT EVENTS.

SOME of the gravest men in America were lately seen in the city of Providence, throwing up their caps as if they would hang them on the horns of the moon. Eye-witnesses say that in Music Hall in that sober municipality there were clappings and shoutings, thumping with canes and umbrellas, stampings with feet, shaking hands, laughter, weeping for joy, waving handkerchiefs, swinging of hats, and in some cases the tossing of them into the air. What was the cause of this demonstration? Simply that a penurious people had paid a debt incurred by penuriousness. [Applause.]. The friends of a most venerable society, which has been known in all zones for fifty years, are proud of having relieved themselves, partly by the aid of secretaries who are statesmen, and who act on democratically small salaries, of a debt that was checking one portion of the advance guard of aggressive religion on benighted foreign shores. Five hundred thousand dollars are to be raised this year, we are told, to strengthen this work at the front; and yet we are assured that no new enterprises can be undertaken with that sum. So penurious is America, that she allows this assurance to be made in face of her opulence, and does not feel ashamed. We have paid a debt which we ought never to have incurred, and we cannot raise money enough to make aggressive advance; and we are loudly congratulating ourselves while we have done painfully less than it is our duty to do.

In the last seventy years the advances of Christianity among those who never heard of it before have been greater than in the first seventy years of the apostolic age. Events not arranged by man have opened all lands to religious truth. Three-fourths of the missionaries under the control of the American Board may be reached by telegraph from Boston within twenty-four hours. There are no foreign shores. Sitting in his office yonder, a statesman secretary with whom I conversed this morning told me that on a Saturday a telegraphic despatch reached him in Boston from a missionary in Japan; and that a reply to it, shot over the wires through England, Germany, Turkey, Asia Minor, India, and China, was received in Japan from Boston the next Tuesday morning; and that a missionary, acting upon intelligence sent thus by

"Thunderless lightnings smiting under seas,"

was then setting sail for America across the Pacific.

Look at the unexplored portions of the world, and you will find that the telegraph is rapidly exploring them; but if a telegraph line can pass through Central Asia, and almost through Central Africa, shall we not send the missionary where commerce carries the electric wire?

The truth is that men underrate the amount that has already been done in Africa. I hold in my hand statistics which show that this darkest of the continents contains, including Madagascar, 130,000 church-members, native born and in mission churches. Five of the vigorous missionary societies of Great Britain are now following up Livingstone to Lakes Tanganyika and Nyanza. Three individuals in the fat land which we recognize as our mother-isle, and which we never have equalled in opulence of gifts to religious enterprises, gave each $25,000 for the purpose of pushing missions in Africa. We have forty millions of people, and Great Britain forty millions. All our missionary societies together collected $1,800,000 in 1875. Those of Great Britain received $3,100,000. In 1875 the American Board collected $468,000; the Baptist Missionary Union, $241,000; the Methodist Episcopal Board, $300,000; the Presbyterian Board, $456,000. But in the same year the Gospel Propagation Society in Great Britain received $400,000; the London Missionary Society, $517,000; the Wesleyan Missionary Society, $500,000; and the Church Missionary Society, $879,000. Our own Stanley is following on the track of Livingstone, and we cannot long consider the interior of Africa as wholly unknown. It is already well enough explored to allow missions to be planted on the lakes discovered by Livingstone. When Stanley shall come back, and show us what Livingstone never saw, will it not be fitting for our different missionary societies to lock hands with each other as those of Great Britain have done; and then to lock hands with hers, and see to it that a permanent beam of light is shot through this last dungeon on our planet?

Long shadows fall from the western mountains of China, and from the Himalayas northward, upon a territory that has hardly yet been reached by Christianity. More than nine-tenths of the population of the Chinese Empire have never heard the central truths of Christian civilization. But Japan is filling with a dawn that will be a Day, and is rapidly crystallizing in the habits demanded by Christian custom. Six thousand towns between the Himalayas and Cape Comorin are Christian. The darkest places are the interior of Africa, the islands between Australia and Asia, and the centre of the Asiatic continent.

How large is the field of the world? Start in the morning at San Francisco by railway, embark eight days later on a steamer at New York or Boston, land at some French port, take the railway to Brindisi, cross the Mediterranean to the Pyramids, and you have travelled eight thousand miles. That is the distance through this little planet. Sometimes I sit in my study, and turn about my globe, and remember that it is no voyage at all to pass from the Golden

Gate to the Pyramids, and yet that this distance is as great as the whole vaunted thickness of the soft-rolling ball on which we wake and sleep. When I look out from the summit of my house-top, and see the watery meridians of the Atlantic dropping downward toward the east until they hide the hulls of the vessels, and leave only thin top-gallants visible, I find it not difficult to bend these aqueous curves in and in around the little space of eight thousand miles until they meet underneath my feet, and I feel the whole globe afloat in the bosom of Omnipotence. This little ball is all home to us. We are to go hence; but while we are here, and looking off into the vast spaces which may be the homes of souls, it is our duty to see that no unexplored remainders are left on this small globe. The iron fingers of commerce are often made to reach around it, as a part of the sport of some merely mercantile enterprises. Why, Lord Bacon shames us, for he says, "Truly merchants themselves shall rise in judgment against the princes and the nobles of Europe; for the merchants have made a great path in the seas, unto the ends of the world, and sent forth ships and fleets of Spanish, English, and Dutch, enough to make China tremble; and all this for pearl, and stones, and spices. But for the pearl of the Kingdom of Heaven, or the stones of the heavenly Jerusalem, or the spices of the Spouse's Garden, not a mast has been set up." God is making commerce his missionary. In this city, and in this audience, are men whose fleets are in all the seas.

It is well known to the closest observers, that it is quite within the power of Christianity to make itself audible by the voice or visible in the printed page, before the end of this century, to every living man.

In the United States in 1776 we had one evangelical minister to every twenty-four hundred of the population: now we have one for every seven hundred. In no other country has Christianity made such outward advance; and to no other land, therefore, are the words more emphatically uttered than to ours, "Preach the Gospel to every creature." Our great cities are listening to tabernacles and to steadily laboring churches. I suppose that there has been as much activity put forth in America to reach the masses at home, as in any other country; but they who work most at home are the most willing to work abroad; and those who are the most willing to work abroad are the most willing to work at home. Echo and re-echo! Those who feel that the field is the world feel also most acutely that their field is their own hearthstone. The reverse is also true. Show me a man who is aggressive in Boston, and I will show you a man who will be aggressive on the Bosphorus, and under the shadows of the Himalayas, and along the rivers of China; who would establish Mount Holyokes and Wellesleys in the South of Africa, and would brave the fevers of the Gold Coast, and carry through the centres of darkness a light such as commerce alone has never diffused, such as only the Bible has shed upon heathendom,—a light which diffuses conscientiousness, and therefore allows property at last to be safely diffused. [Applause.]

If, from a visible throne in the heavens, He whom we dare not name were to send a troop of angels to the centre of Africa, and another to the interior of Asia, and another to Japan, and another to the isles of the Pacific, and if, by the activity of these visitants, there should be broken open a way for commerce in Japan, a way for missions in China, a way for religious truth in the centre of Africa, we should all bow down and adore before such a revelation of the purposes of Providence. But a Power not of man has sent visitants to Japan, and to the isles of the sea, and to the centre of Asia, and the heart of Africa. Treaties with once rusty hinges, whose turning grated sounds of war, now move as if all their joints were oiled. Bulwarks of ages have fallen down. The interiors of continents not long ago largely unknown to geography are open at this hour to missions. These events are just as surely the results of Divine Providence as if they had been brought about by bands of heavenly visitants. It does not become us to exhibit elation because we have treated Providence penuriously, and at last have paid the debts into which we fell by lagging behind Almighty God. We are not to be ashamed of missions, for God evidently is not ashamed of them. [Applause.]

THE LECTURE.

In 1786 Frederick the Great lay dying at Sans Souci; and in 1865 Thomas Carlyle, face to face with all the scepticism and doctrinal unrest and small philosophy of our time, and with a mind free as Boreas horsed on the north wind, sat down to describe the scene of Frederick's departure. This all-doubting man Frederick, a pupil of Voltaire, seemed to have neither fear nor hope in death; but, says Carlyle, there was one kind of scepticism which he never could endure. "Atheism, truly, he could not abide. To him, as to all of us, it was flatly inconceivable that intellect, moral emotion, could have been put into him by an Entity that had none of its own." (*Life of Frederick*, vol. vi., last chapter.) Carlyle affirms that to *all* of us it is inconceivable, and this *flatly*, that evolution can exceed involution; or, that we can have intellect, emotion, conscience, as the gifts of a Power that has itself none of these to give.

You remember, gentlemen, that Webster, when asked what his greatest thought was, looked about on the company at a crowded dinner-table, and asked, "Who are here?"—"Only your friends."—"The greatest thought that ever entered my mind was that of my personal responsibility to a personal God." He expanded that idea in conversation for ten minutes, and rose and left the table. Men stood and sat in the hushed room, saying to each other, "Did you ever hear any thing like that?" But yonder, on the shore of the sea, this same Webster, closing the greatest legal argument of his life,—a document which I now hold in my hands,—uttered the same thought in words that I have read standing on the coast there, and which have in them, whether read there or here, or anywhere on this lonely shore of existence, which we call life, a giant swell like the roll of the Atlantic, an instinctive colossal tide found in every soul that is possessed of the full equipment of a man. "There is no evil that we cannot either face or flee from, but the consciousness of duty disregarded. A sense of duty pursues us ever. It is omnipresent like the Deity. If we take to ourselves the wings of the morning, and dwell in the uttermost parts of the sea, duty performed, or duty violated, is still with us, for our happiness or our misery. If we say that darkness shall cover us, in the darkness, as in the light, our obligations are yet with us. We cannot escape their power, nor fly from their presence. They are with us in this life, will be with us at its close; and, in that scene of inconceivable solemnity which lies yet farther onward, we shall still find ourselves surrounded by the consciousness of duty, to pain us wherever it has been violated, and to console us so far as God has given us grace to perform it." (*Webster's Works*, vol. vi. p. 105.)

Flatly inconceivable that moral emotion, intellect, can have been put into us by a Being that has none of its own! But Matthew Arnold says that neither this inconceivability, nor any thing else, shows that God is a person. It is a physical fact that Matthew Arnold's upper forehead is very flat. Here are Car lyle, Frederick the Great, Webster; and I might put with them Cicero, Plato, Aristotle, Leibnitz, Kant, Richter. Indeed, the latter says, speaking from experience, and for men of his own natural rank, that the summit of every full-orbed nature suggests the belief in God as a person. At the top of the great hills in Italy, we commonly find chapels. Richter affirms (*Titan*) that in the heights of every fully endowed man, there is an instinct of obligation, or sense of responsibility, which points to a personal God. So Schleiermacher said, and built a renowned and to-day not uninfluential system of religious thought upon the assertion; but he was a theologian. So Kant taught in his theory of the practical reason; and German philosophy at the present hour, however shy of some of his outworks, dares build nowhere else than on his fundamental principles; but he was an ethical philosopher. Take only literary men, take lawyers, take historians, take philosophers of no school in ethics, and, as a general and very revelatory rule, wherever they have been full-orbed, they have found in the depths of their endowments, this deepest instinct, — a sense of obligation, a feeling of dependence.

> "Below the surface stream, shallow and light,
> Of what we say we feel; below the stream,
> As light, of what we think we feel, — there flows,
> With noiseless current strong, obscure, and deep,
> The central stream of what we feel indeed."

A highly important question in our vexed time is whether we are to take for our general guides men possessing the full range of natural endowments, or fragments of men, brilliant, indeed, in parts of the human equipment, but lacking several things that go to make up a full-orbed man. I am not here to assail any person as naturally unequipped. But we are most of us fragments; and Mr. Arnold admits, and his critics have always insisted, that among his limitations is a great deficiency of metaphysical insight. "Men of philosophical talents will remind us of the truths of mathematics," says Matthew Arnold himself, "and tell us that the three angles of a triangle are undoubtedly equal to two right angles; yet, very likely from want of skill or practice in abstract reasoning, *we cannot see the force of that proposition*, and it may simply have no meaning for us. The proposition is a deduction from certain elementary truths, and the deduction is too long or too hard for us to follow; or, at any rate, we may have not followed it, or we may have forgotten it, and therefore we do not feel the force of the proposition." " Here it is, we suppose, that one's want of talent for abstract reasoning makes itself so lamentably felt." (*God and the Bible*, pp. 69, 70; London, 1875.) "Probably this limited character of our doubting arose from our want of philosophy and philosophical principles, which is so notorious, and which is so often and so uncharitably cast in our teeth." (*Ibid*, p. 62.) "We are so notoriously deficient in talents for metaphysical speculation and abstruse reasoning, that our adversaries often taunt us with it, and have held us up to public ridicule, as being without a system of philosophy based on principles interdependent, subordinate, and coherent." (*Literature and Dogma*, p. 389; London, fourth edition, 1874.) Matthew Arnold admits that all metaphysics are to him "the science of non-naturals." (*God and the Bible*, p. 50.) But by metaphysics we understand here, as people do elsewhere, the science of self-evident truth, — a systematic examination of axioms, with the inferences that all men must draw from them, if they are only true to the self-evident propositions which all admit. Metaphysics may, indeed, be so treated as to be obscure; but metaphysics rightly treated is the luminous and exact science of self-evident truth. Matthew Arnold flaunts it as a science of non-naturals; and, because some proof of the existence of God is drawn from metaphysics, he will have nothing to do with any conclusion that stands on this pedestal, — an abstract, all in the air, as he calls it.

Incontrovertibly we do not stand on any thing that rests in the air when we stand on these ineradicable human instincts which belong to every full-orbed nature, — a feeling of dependence, a feeling of obligation. Each is a part of us. We are so made that we cannot doubt our finiteness. We are not everywhere; we do not possess all power. There are limitations of our being. But we have an idea of the Infinite. We are circumscribed, and we have an idea of a Being who is not. We do not comprehend him, but we apprehend him. As individuals we began to be. There is evidence that our race began to be. Once man was not on the globe; he came into existence. Whatever begins to be must have a cause. We cannot suppose that the Infinite has come forth from the finite. We, the caused finite, must be the work of the Infinite. In loyalty to self-evident truth, we must put the finite in the relation of effect, and the Infinite in the relation of cause; and so we begin to feel sure, in the name of all clearness of thought, that we can intellectually justify this instructive sense of dependence.

There is an Eternal Power, not ourselves, on whom we are dependent; this is, indeed, Arnold's central thought. Nothing is more beautiful in his writings than the steady melody of one chord in his harp. Most of the chords are too short, or twisted, or unduly strained; but there is one note in Matthew Arnold which has a divine resonance, and that is his passionate preception and proclamation of the natural victoriousness of right under the laws of the universe. Everywhere he is the prophet of a Power, not ourselves, which makes for righteousness; and this central assertion of his he regards as a truth of absolute science. He cannot decide whether the Power is personal or not. He will not deny that it is a person. The Edinburgh Review says to him, "All existing things must be persons or things. Persons are superior to things. Do you mean to call God a thing?" Matthew Arnold replies, "We neither affirm God to be a person nor to be a thing. We are not at all in a position to affirm God to be the one or the other. All we can really say of our object of thought is that it *operates*." (*God and the*

Bible, pp. 97, 98.) There is in the universe an Eternal Power which makes for righteousness. We know this, as we know that fire burns by putting our hands into the flame. It is not necessary to decide whether this power is or is not a person. I know by its operation on me, by its influence in universal history, by the instincts which point it out, and by my sense of personal dependence and obligation, that it makes for righteousness.

Standing now on this common ground, I wish to lead you up the heights which rise from it; and, whether Matthew Arnold accompany us or not, I know that others will, — the Kants and the Schleiermachers, and the Richters, and the Ciceros, the Platos and the Carlyles and even the Fredericks the Great; and thus, if we go up without Matthew Arnold, we shall not go up in bad company. [Applause.]

1. Conscience emphasizes the word ought.

2. That word expresses the natural, human, instinctive sense of obligation to moral law.

3. It is everywhere admitted that this law was not enacted, and that it is not reversible by the human will.

4. It is imposed on us by an authority outside of ourselves.

5. Our obligation is, therefore, to an authority outside of ourselves.

6. *Our instinct of obligation is active even when we are separated from all human government and society.*

7. *We cannot imagine ourselves to obliterate the dis-*

tinction between right and wrong, even by the obliteration of all finite beings and of all immaterial nature.

I can imagine the putting out of all the fires of all the hosts of heaven. I can imagine that all finite being here and in the Unseen Holy is not. But I cannot suppose that the putting out of existence of all finite being would obliterate the distinction between upper and under, between the whole and a part, between a cause and an effect, or between right and wrong. The difference between the right hand and left would yet inhere in the very nature of things, were all finite existence swept out of the universe. It would yet be true that there cannot be a before, without an after, that two straight lines cannot enclose a space, and that there is a difference between the whole and a part, and between right and wrong. These propositions are self-evident truths, and depend for their validity, not on the existence of the archangels, or of the government of the United States, or of Magna Charta, or of the human race. They are revelations of the laws of the nature of things, existing before Rome was founded, and, as Cicero used to say, likely to retain their authority when all human empires have been swept away. It is a very strategic point that I am elaborating; but I believe, now that I ask you to judge for yourselves, that I carry your general assent in asserting that we may imagine the annihilation of all finite existence, and yet, after that, have the existence of a distinction between the whole and a part, between a cause and an effect, and between right and wrong. This latter distinction, however, is only another name for the moral law; and so Webster is right. The sense of duty pursues us ever. Even when these visible heavens are rolled away, the moral constellations remain, and pursue their accustomed courses in the invisible heavens which never shall be rolled away.

8. On examination of personal consciousness it is found, therefore, that this authority to which we owe obligation is not immaterial nature, not the human-race, not human government and society, nor finite being in general.

All these things we can imagine annihilated, and yet our sense of duty pursues us ever. The feeling of obligation, that is, of the difference between right and wrong, and that the right ought to be chosen and that the wrong ought not, continues to follow us.

9. We know through conscience that we must answer for what we are, and for what we do, to a Power outside of us.

10. *In the very nature of things, moral obligation to answer for ourselves to a Power not ourselves can be owed only to a Power that knows what we are and what we do, and what we ought to do; who approves of the right, and disapproves of the wrong; and who has the power and purpose to punish or reward us according to our character and conduct.*

11. Such being the facts of our moral nature, we are under the necessity of assuming the existence of such a Being or Power, by whatever name we call it.

12. Such a Being or Power, who knows what we are and what we do, and what we ought to be and

do, and who approves of the right, and disapproves of the wrong, and who has the power and purpose to punish or reward us according to our character and conduct,—such a Being or Power is a personal God on whom we are dependent, and to whom we owe obligation. [Applause.]

This is the argument by which Kant and Hamilton, while denying the validity of all other arguments for the existence of God, are forced to admit that our nature compels us to believe that He is, and that He is a Person. Probably this argument, which convinces scholars more than any other, is the one which convinces the mass of men more effectively than any other form of reasoning from the organic instincts of conscience.

Some men hold, and I will say nothing against their reputation for scholarship, that the existence of God is an intuition, or that we know that He exists just as we know that every change must have a cause, or that a whole is greater than a part. I, as you are already aware, do not hold that the Divine existence is guaranteed to us by intuition. It is evident, but not self-evident. It is guaranteed to us by a single step of inference, from our deepest, surest, most ineradicable instincts. When I analyze these, I find the fact of God's existence as a Person lying capsulate, wrapped up in the sense of dependence and of obligation, which *are* intuitions. I am just as sure that I am a dependent being as I am that two and two make four. I am just as sure that I am under obligation to what ought to be, as I am that a whole is greater than a part.

The difference between right and wrong in our choices and intentions, you will find to be not only evident, but self-evident. You will allow me here and now, since I do not say the Divine existence is guaranteed to us by intuition, to affirm that the distinction between right and wrong is thus guaranteed. That there is a distinction between right and wrong in choices, is beyond all controversy, just as it is beyond all controversy that the whole is greater than a part. One of these assertions is as self-evident as the other. When we perceive this distinction between moral motives, we feel that we ought to obey a good motive, and disobey a bad. Thus our sense of obligation expressed by the word *ought* is guaranteed by intuition as well as by instinct. Intuition stands on one side of it, and instinct on the other. The feeling that we ought to obey the right motive is the instinct; the perception of the right motive is the intuition. Conscience perceives the distinction between right and wrong in choices, and feels that the right ought to be performed, and that the wrong ought not to be. Thus direct intuition and organic instinct, the two highest authorities known to man, guarantee to us this sense of dependence and this sense of obligation. In the study of conscience we stand between the two pillars on which all surety rests; and, looking upward along the flutings of these two shafts of intuition and instinct, — perception of the difference between right and wrong in moral motives, and feeling that the right ought to be followed and that the wrong ought not, — we can throw an arch from the capital of one shaft to that of the other, and on its summit, the sense of dependence on the one side and the sense of obligation on the other, we place upon the keystone the lowermost corner of the house not built with hands, the belief in a personal God to whom we owe that obligation, and on whom we are thus dependent. [Applause.]

If, however, you refuse, with Matthew Arnold, to examine self-evident truths as a science, I must ask you to take the point of view of the microscope. Here is a course of thought proceeding out of the very heart of Biology: —

1. Some force forms the parts in an embryo. "We are woven," even Tyndall says, "by a power not ourselves."

On the 1st of October, at the Midland Institute, Professor Tyndall gave to the world knowledge of a secret which most scholars have understood for ten years. At the Midland Institute, in that city of Birmingham, which is so well known to you, sir (turning to the Rev. Dr. Dale of England), Professor Tyndall said to the robber, the ravisher, and the murderer, "You offend because you cannot help offending." (Report in London "Times" of Tyndall's lecture of Oct 1.) Häckel affirmed years ago, in his History of Creation (vol. i. p. 237), that "the will is never free." Some of you have thought it extravagant to assert that this same teaching lies between the lines of many a page published by the English materialistic school. Häckel is far bolder than most of his followers, and he has proclaimed pointedly that the will is never free; and now Tyndall does the same. With much grace, with high literary ability, and with all the prestige of his great name, Professor Tyndall says to the murderer, "You offend because you cannot help offending; we punish you because we cannot help punishing." Approbation and disapprobation he would no more have as to the overflow of the muddy torrent we call an Iago or a Mephistopheles than he would for the overflow of the Rhine or the Mississippi. According to his scheme of thought, we may put up dykes against Caligula and Nero as we do against the Mississippi, but we are not to have disapprobation for Caligula, or for Nero, or for Catiline, any more than for the Tiber when it overflows its banks into the marble temples of Rome. We must say to the criminal, "You offend because you cannot help offending." These are Tyndall's own words, which Hermann Lotze would think hardly worthy of a reply. They are not more penetratingly mischievous than violently unscientific.

But even Tyndall asserts that we are woven by something not ourselves. (Lecture at Birmingham, Oct. 1.) Now, I affirm that when the embryo comes into existence, some force forms its parts. The force that forms the parts is the cause of the form of the parts. The cause must exist before the effect. We are sure of that, are we not? My delicious and surprising friends, who are sure of nothing except that you are sure you are sure of nothing, thereby contradicting yourselves, are you not certain that a cause must exist before a change can be produced?

Very well: here I stand, with the process of the weaving of a physical organism going on under my microscope. Here is woven a lion, there a man; here an oak, there a palm. From the first the plan of each is in the embryo from which each begins. That plan must have been in existence before any physical organization exists in the embryo. Even your Häckel says ("Popular Science Monthly," October, 1877, article on Bathybius, p. 652), that " Life is not a result of organization, but *vice versa.*" It is demonstrable under the microscope, that life is the cause of organization, and not organization the cause of life. The plan must be in existence before it is executed. A plan in existence and not executed is a thought. The plan executed in the weaving of an organism, therefore, was a thought before the organism was woven. That thought exists before the organism. But thought implies a thinker. There cannot be a thought without a thinker. The thought executed in the organism does not belong to the organism. The design is not in the thing designed: it is outside the thing designed. The cause is outside of the effect. Thought, the force that forms the embryo, is not in the embryo: it is outside the embryo, for it exists before the embryo. Talk as you please about force being inherent in all matter ; or of the tree Igdrasil, as Tyndall has lately said, being the proper symbol of the universe : we know that the cause must exist before the change it produces. This plan by which the form of the embryo is determined must be in existence somewhere before any form is woven. The first stroke of the shuttle, as we have proved, implies a plan; and so we know that there is in the universe a thought, not ourselves and not our own. Adhere to that proposition, and use Descartes' great argument, — " I think: therefore I am a person."

2. Since we are woven by a power not ourselves, there is thought in the universe not our own.

3. There cannot be thought without a thinker.

4. Therefore there is in the universe a thinker not ourselves.

5. But a thinker is a person. [Applause.]

To put now the whole argument from design into the shape which best pleased John Stuart Mill, we may say: —

1. Every change must have an adequate cause.

2. My coming into existence as a mind, free-will, and conscience, was a change.

3. That change requires a cause adequate to account for the existence of mind, free-will, and conscience.

4. Involution must equal evolution.

5. Only mind, free-will and conscience, in the cause, therefore, are sufficient to account for mind, free-will, and conscience in the change.

6. The cause, therefore, possessed mind, free-will, and conscience.

7. The union of mind, free-will, and conscience in any being constitutes personality in that being.

8. The cause, therefore, which brought me into existence as a mind, free-will, and conscience, was a person.

If you will look at that list of propositions, you will find nothing taken for granted in them except that every change must have an adequate cause. I suppose them to be substantially the ground on which established science stands to this hour, with the Richters, and the Carlyles, and Platos, and Aristotles, and even with the all-doubting Fredericks.

We may say also, in presenting further the argument from design: —

1. If there is an omnipresent, self-existing, and infinitely holy moral law, and if the nature of all dependent intelligence has been adapted to that law, there must be a moral designer to account for this moral adaptation.

2. There are such a law and such an adaptation.

3. There is, therefore, a moral designer.

4. But a moral designer must possess mind, free-will, and conscience.

5. The union of mind, free-will, and conscience in any being constitutes personality in that being.

6. The moral designer of the moral law is, therefore, a person.

John Stuart Mill advised all who would prove the Divine Existence to adhere to the argument from design. Even Matthew Arnold says that all he can say against the argument from design is, that he has had no experience in world-building. "We know from experience that men make watches, and bees make honeycombs. We do not know from experience that a Creator of all things makes ears and buds." (*God and the Bible*, pp. 102, 103.) What if Red Cloud and Chief Joseph had been brought to the Centennial or to Washington? What if they had seen the majestic dome of our national Capitol, and all the marvels of the Centennial? Red Cloud would have said, if he had followed Matthew Arnold's philosophy, "I have had experience in building wigwams. I know the path from my house to the hut of Seven Thunders or Bear Paw. I know that every such path is made by some cause. I know that every wigwam must have been built by some man. But this railroad, — I never had experience in building railroads, — I do not know but that it was fished out of the sea. This marble Capitol, these wonderful and strange things in the Centennial! I have never had any experience in making columbiads or spinning-jennies. I know that the flint which I sharpen for my arrow must be shaped by some man; but this columbiad, I do not know but that it grew. This spinning-jenny! I have had no experience in factories and weaving-machines and these marvels. I think this loom was *evolved!*" [Applause.]

I do not in the slightest degree misrepresent the reasoning of Mr. Arnold; for the only objection he has to the argument from design is that he has had no experience in world-building. David Hume also once made that assertion; but when he walked with Adam Ferguson on the heights of Edinburgh one night, and studied the constellations, he said, "Adam, there is a God."

Stuart Mill admits that the argument from design proves the existence of a designer; but whether we can prove that the designer thus proved to exist is the only designer in the universe, is, as some people think, yet left in doubt. Paley's argument is supposed to be overthrown. A watch implies a watchmaker; but how do we know that there was not a designer of the watchmaker, or that there is not a second God that designed the first God, and a third that designed the second, and so on? A design must have had a designer, and the designer a designer, and this designer a designer; for every design is to have a designer. Do not suppose that I am here to dodge this difficulty, although occasionally it may be that some of our theological teachers have evaded it. I have heard that Lyman Beecher was once approached by his students with the question how they should answer sceptics who told them that the argument from design proved too much. "They say to us," the students told their teacher, "that there may be twenty Gods, for every design must have a designer, and every designer a designer, and so on." Now, Lyman Beecher did not know how to answer that difficulty, or at least he did not give the scientific answer; but he was quick in thought, and so he said to his students, "These men say there are twenty Gods?"—"Yes."—"Well, you tell them that if there is one God it will go hard with them, and if there are twenty it will go harder yet." [Applause.] But the answer to be made is that we cannot have a dependent existence without an independent or a self-existent being to depend upon. All existence, to put the argument in syllogistic form, is either dependent or independent. You are sure of that? Yes. Well, if there is a dependent existence, there must be an independent, for there cannot be dependence without something to depend upon; and an infinite series of links receding forever is an effect without a cause. Your axiom that every change must have an adequate cause is denied by the theory of an infinite series. You carry up your chain, link after link, and there is nothing to hang the last link upon.

1. All possible existence is either dependent or independent.

2. If there is dependent existence there must be independent existence, for there cannot be dependence without dependence on something; an endless chain without a point of support is an effect without a cause; dependence without independence is a contradiction in terms.

3. I am a dependent existence.

4. Therefore there is independent existence. [Applause.]

But independent existence is self-existence.

1. All possible being is either self-existent or not self-existent.

2. If there is being which is not self-existent, the principle that every change must have an adequate cause requires that there should exist being that is self-existent.

3. I am a being that is not self-existent.

4. Therefore there is being that is self-existent.

So, too, with exact loyalty to self-evident truth, we may say:—

1. All possible persons are either self-existent or not self-existent.

2. If there exist a person that is not self-existent, there must be a person that is self-existent.

3. I am a person not self-existent.

4. Therefore there is a Person who is self-existent. This is He.

In these arguments nothing is assumed but self-evident truths, which all men act upon in business, and take as certain at the fireside. The deep human instincts of conscience proclaim all that our metaphysics do. Science, standing upon axioms, knows no more at last than the man full-orbed, who allows every tide in him to rise according to untaught instinct, and finds that when he swells aloft under the natural attraction felt by the sense of obligation and dependence, he touches the stars. If you are a thin brook; if you are under the torrid sun of scepticism; if there are no great waves in you that can kiss the heavens at times, — you may be in doubt. But let your nature become oceanic, and feel all that can come to you from the winds, and from the springs, and from the search of the depths; and then, when the Power, not ourselves, that makes for righteousness, rides the waves, you will find that the highest instincts in you touch Him far aloft, as a Person. [Applause.]

V.

ORGANIC INSTINCTS IN CONSCIENCE.

THE EIGHTY-FIFTH LECTURE IN THE BOSTON MONDAY LECTURESHIP, DELIVERED IN TREMONT TEMPLE, OCT. 29.

The Moral Sense, thank God, is a thing you never will account for; that, if you could think of it, is the perennial Miracle of Man; in all times visibly connecting poor transitory man here on this bewildered earth with his Maker who is eternal in the Heavens.
CARLYLE: *Shooting Niagara: and after?* vi.

Das Gewissen ist das Organ zur Manifestation der göttlichen Gerechtigkeit im menschlichen Selbstbewusstsein.
HOFMANN: *Das Gewissen*, ii.

V.

ORGANIC INSTINCTS IN CONSCIENCE.

PRELUDE ON CURRENT EVENTS.

SUPPOSE that there should be called into existence in the Eastern or Western States a million voters unable to read. Were the Northern portion of the Union suddenly saddled with danger of this kind, our vigor would bestir itself, no doubt, to shake off the incubus of so large a mass of enfranchised ignorance. But the Southern States have had brought into their borders lately, by an act of our General Government, one million voters unable to read. The population of the territory which we call the South is slightly larger than that of the Eastern and Middle States, or than that of the section which we call the West. Whether you approve the policy of the Chief Executive of this nation or not, it is one of pacification. We placed the flat side of the sword on the neck of the South for a while, after the keen edge had caused her to surrender. We kept the flat edge of the bayonet on her neck in order to secure peace at elections, and peace for the freedman's lonely school on the edge of the Dismal Swamp, and peace for all unarmed men at night. We did not always secure what we wished. But now the flat side of the bayonet and of the sword has been taken off. There is no method of managing enfranchised ignorance at the South, except by educating the freedmen.

It would be a felicity if this audience could assemble in imagination in some freedman's solitary schoolhouse in the Florida Everglades, or under the moss-hung pines of the Carolinas and Mississippi, and meditate there a moment on the duties of the North toward uneducated voters created by its own act. Once in personal contact with the South, we find a strange land, fat, semi-tropical in places, capable of great wealth, but many old plantations are covered with weeds. If the Confederate soldiers in their graves could come back, they would find not a few of their old homes unrecognizable. Capital was greatly centralized by slavery, and now it is being decentralized. The only prosperous portions of the new South are the regions where men have started small farms, and operate them upon the principle that machinery is to be used and labor paid for. In cases where small farms have decentralized capital, prosperity is slowly returning to the Gulf States. Even in those quarters of the South where free labor is thus tardily acquiring honor, the negro is in debt. He is paid for his labor, but he is in debt at the country store. Authorities exceptionally well acquainted in the South assert that these debts at the corner groceries are carefully fostered. The freedman does not easily buy land in the South. The citizen who was lately a slave is paid very little for his labor, and falls into debt. He cannot leave the farm on which he lives, unless his debts are cancelled. It is the scheme of many an old master, that these debts shall not be too swiftly paid. Put your ear to the ground in some of the best society in the Middle States, and you will find not a little tremor there from the fear that a time may come within fifty years when a large part of the black race will fall into the condition of the Mexican peons, held in a kind of qualified bondage for debt. If you do not anticipate trouble from that source, it is yet certain that many do; and I cannot undertake to assert, at this distance from the scene, that there is not a threat in that cloud which lies half out of sight along the Southern horizon. [Applause.]

But there are much blacker clouds there. A strange land this, over the mellow acres of which we gaze from the windows of our freedman's schoolhouse. Twenty-five and five-tenths per cent of the population of the South over ten years of age cannot read. Thirteen millions are here, and a quarter of them need to use the spelling-book yet. I speak with all sympathy for a section of our nation which has exhibited great bravery, and is certainly able to educate its citizens if it has the will to do so. But it has not had that will. At this moment, it is true that my native State of New York spends more for education than all the South. If your uneducated freedmen were as well educated as the average

Southern white men, they would not be well enough educated to take care of themselves, and become intelligent voters. Thirty-nine per cent of the voters of the South cannot read the names of the candidates printed on their ballots. Three and eight-tenths per cent of the Middle States and New England are illiterate, — that is, of the population over ten years of age, that percentage cannot read. Three and four-tenths per cent only of the Western States are illiterate. I whisper this in Boston: we are behind the upper part of the Mississippi valley. In Alabama fifty-three per cent of the voters are illiterate; even in Kentucky twenty-eight per cent are illiterate; in Maryland twenty-two; in Delaware twenty-four. Of the 2,000,000 illiterate voters in the United States, 1,700,000 are in the Southern States, which elect 32 of the 74 senators and 109 of the 292 representatives in Congress.

Here is a mass of uneducated suffrage, and who is exploiting it? Look at the negro in his schoolhouse. Behind him is his master, to whom he is in debt; and on the other side of him is a strange figure in American politics, not often seen in our land, but one that has been potent in the politics of other lands. This historic form wears ecclesiastical robes. I open authentic documents concerning the condition of the freedmen, and find them resolving the other day, in a grave public assembly at Macon: "That this meeting appoint a committee to wait upon the Rt. Rev. Bishop Gross, who is now in this city, to obtain his views as to the educational policy of the Catholic Church in regard to the colored people of the South, and to ascertain to what extent we may look to that organization for assistance in the work of educating our children." Other documents assure us that from Baltimore there has lately been projected a great aggressive campaign upon the South. New schools for colored children are to be immediately opened, ten in Georgia, fifteen in Alabama, twenty-five in Louisiana. These Romish schools will offer board and tuition free to colored young men and women. If the uneducated suffrage of North and South in one mass is ever to be exploited by a single hand on the Tiber, a serious hour is ahead of us.

Rolling through the Berkshire hills, a few days since, and up the fat valley of the Mohawk to Syracuse, to address an audience for the purpose of arousing interest in the efforts of Protestant free schools in the South, I studied on the way the case of the six thousand pupils of these struggling, heroic institutions at Nashville and Atlanta and Talledega and Memphis. So great was the contrast of their poverty with the opulence of the Connecticut, the Hudson, and the Mohawk valleys I glided through, that I found myself growing sick at heart as I looked out of the car-windows. Schools for freedmen in the South depend yet almost exclusively upon the North for their support. No doubt the freedmen help themselves as far as they can, but they are exceedingly poor. There are men who wish to teach their brethren, both in secular things and divine; and they are burning pine-knots instead of candles, for they cannot pay for the latter. They wish to go out five, ten, fifteen miles into the country, and cannot pay their railway fares; and so for any distances under twenty miles they walk. Again and again their lonely visits in the country-side are subjected to insults from roughs of the poor white class. A negro preacher is not a welcome guest at a planter's mansion. It is only yesterday that the South had in it armed bands which often prevented negroes from voting. Freedmen's schoolhouses, including churches used as schoolhouses, have now and then been burned. Whole tiers of counties were subjected to political terrorism. No doubt the negro has made mistakes. He had a majority in Mississippi; and he did not act there like a saint, but very like an uneducated black rascal. He did things in his official capacity to which I would not have submitted, had I been a citizen in that State; but he acted as it was to have been expected that he would, without education, and with slavery behind him. In South Carolina, the black man has a majority; and he has not acted there like a citizen understanding his duties, but like an uneducated freedman. He has gone to the wall in Mississippi, in spite of being in a majority. He will go to the wall in South Carolina, in spite of being in a majority there. If you would keep him from being pressed to flatness against that wall, you must do so by ringing his school and college bells.

In the cause of the freedman, the bugles of Gettysburg were once at the front; the bugles of Antietam, the bugles of Lookout Mountain. But to-day in that cause, as holy now as ever, the only sounds we hear at the front are affrighted, half-choked noises of school and college bells. [Applause.] You who answered the bugles at Antietam, you who answered the roll of the drums in the smoke of Lookout Mountain, you who understand how many unknown graves there are in the South, will you not hear the confused noise of the freedmen's college bells, and follow them with righteous and victorious aid, as once you followed the bugles? [Applause.]

THE LECTURE.

Plato used to say that a ship is all but its wood. The eloquent shaft on Bunker Hill yonder is not fully analyzed by us when we take into view only its granite. The various parts together exhibit a plan; but all the parts taken separately, and without that plan, are not the monument. The parts of any mechanism without their plan are not equal to the whole. Here is the human eye, or ear, or hand; and each contains more than the sum of all its visible parts. We know that the eye consists of several distinct portions; and when these and their collocations are examined separately we find that they have only one thing in common, namely, the fitness to produce, when each part is co-ordinated with the rest, the organ of sight. We have lenses; we have aqueous and vitreous humors; we have eyelashes; we have the iris; we have the miraculous retina; and, if these were seen in separation from each other, we might at first be unable to find any similarity between them. The retina is not like the crystalline lenses. The substance of which the iris is composed is in great part very different from that of which the lashes consist. Nevertheless, when we study the parts more minutely we find that they have one thing in common,— an adaptation to be a part of a multiplex whole, constituting an organ of sight. Now, that common element in them all *is* something, if you please. It must not be overlooked by the scientific method. There exists undeniably a common element in all the parts of the eye and in their collocations, and it must have had an adequate cause. When all the parts are put together, they constitute an organ of sight; but that sight itself does not spring up until the parts are put together. If the shape of any one part be changed materially, or its collocation altered, sight ceases or is impaired. Every part has such a relation to the whole, that each harmonizes with all the rest in an adaptation to produce an organ of sight; and so we feel sure that the adequate cause of that adaptation must have had in view sight as the result of this one common element in all the portions of the eye. The only adequate cause is something that intended to produce sight at the end of the process which brought into existence these parts and their arrangement.

Whether the parts came together by evolution, or by special creation; whether God's will operated through unchanging laws or by a special act, to produce the eye, we know that somewhere this adaptation of each part to the one aptitude of the whole mechanism must have had a sufficient cause. Even John Stuart Mill, sceptic as he was on many points, admits explicitly that we cannot explain the adaptation of part to part in the eye without supposing that the idea of sight goes before the adaptation of these pieces to each other in such a manner as to produce sight. There must be an idea before we can have a plan; and here an idea plainly existed before the effects it produces. The effects are the various parts of the eye and their adaptation to sight; but sight starts up only at the end of a long process. The idea of sight as an end to be attained must have been in existence somewhere, when the adaptation of piece to piece was secured. That idea we prove to exist, not by analogy merely, but by induction. "This," Mill says in his last book (*Three Essays on Religion*, American edition, pp. 171, 172), "I conceive to be a legitimate inductive inference. Sight, being a fact not precedent but subsequent to the putting together of the organic structure of the eye, can only be connected with the production of that structure in the character of a final, not an efficient cause; that is, it is not sight itself, but an antecedent idea of it, that must be the efficient cause. But this at once marks the origin as proceeding from an intelligent will." This logician makes this last stupendous concession, because he knows very well that there cannot be an idea without a mind to contain it. There cannot be a thought without a thinker, any more than there can be an upper without an under, a before without an after, a here without a there. Reasoning, therefore, upon the strictest principles of inductive logic, applying all the tests of the scientific method, Stuart Mill's conclusion is that an antecedent Idea of sight must be the cause of sight, and that this Idea must have existed in a Being possessing an intelligent Will. [Applause.]

Herbert Spencer very inexcusably mistakes the force of such reasoning as this of Mill's, and calls it the carpenter theory of the universe. Spencer's own scheme of thought, involving implicitly, as Häckel's does explicitly, the assertion that organisms have come into existence by spontaneous generation or fortuitous concourse of atoms, shaken about like dice in a dicer's box, I call the dicer's theory of the universe. For one, I prefer the carpenter theory to the dicer's theory; but I hold neither the one nor the other. Mill discusses the dicer's theory, and is of course candid enough to admit that "this principle does not pretend to account for the *commencement* of sensation, or of animal or vegetable life." He weighs all his syllables, and commits himself and his philosophical reputation in the last year of his life to the proposition that "it must be allowed that the adaptations in nature afford a large balance of probability in favor of creation by Intelligence." (*Three Essays on Religion*, p. 174.) "The number of instances [of such adaptations] is immeasurably greater than is, by the principles of inductive logic, required for the exclusion of a random concurrence of independent causes, or, speaking technically, for the elimination of chance." (*Ibid.*, p. 171.) Thus Herbert Spencer failed to convert the last of the world's great logicians to the dicer's theory so dear to all materialistic schools of thought. Scientific Theism holds neither the carpenter theory nor the dicer's theory of the origin of the universe, but asserts Goethe's proposition:—

> "Who of the living seeks to know and tell,
> Strives first the living Spirit to expel, —
> He has in hand the *separate parts* alone,
> But lacks the spirit-bond that makes them one."

It is the supreme principle of Herbert Spencer's philosophy, as well as of Sir William Hamilton's, that any thing we cannot help believing, or any proposition of which the opposite is utterly inconceivable, we must hold to be true. This has been the fundamental principle of every philosopher worthy of the name since Aristotle. Utter inconceivability, I claim, inheres in the proposition that this adaptation of part to part in the eye can be produced without the preceding idea of sight. Utter inconceivability lies behind all atheistic thought. So too, it lies behind all thought which does not deny that God exists, but denies that we can know that he does. This agnostic theory never makes a scientific use of axioms; it denies the power that inheres in necessary beliefs; asserts with Spencer that we must consider as true our necessary beliefs; and then with him denies that these beliefs carry us out to the idea of an Intelligent or Personal First Cause.

Thus far I have endeavored to lead you through this lecture, as through the last, over the ground of Induction, based upon Intuition. But, to turn now to the ground occupied by the great Organic Instincts of Conscience, it is not uncommon to find even materialism admitting that men instinctively think of God as Personal. It is often conceded that our instincts point that way, but we are assured that our instincts mislead us. We have been miseducated. There are lying faculties in us. Our profoundest tendencies raise false expectations. It is on this verge of the wildest kind of scepticism, on this edge of what the schools call Pyrrhonism, on this border of the denial of all self-evident truth, that I wish to call pause to-day for a moment, in the name of the axioms of science.

Here is the best book on the scientific method that has been produced since the death of Sir William Hamilton. You will all allow me to say that the Principles of Science, by Professor Stanley Jevons, is a standard work; but he closes his hundreds of pages, filled with the most careful analysis of logical forms, with these very incisive sentences: " Among the most unquestionable rules of Scientific Method is that first law, that whatever phenomenon is, is. We must ignore no existence whatever; we may variously interpret or explain its meaning and origin, but if a phenomenon does exist, it demands some kind of an explanation. If men do act, feel, and live, as if they were not merely the brief products of a casual conjunction of atoms, but the instruments of a far-reaching purpose, are we to record all other phenomena, and pass over these? We investigate the instincts of the ant and the bee and the beaver, and discover that they are led by an inscrutable agency to work toward a distant purpose. Let us be faithful to our scientific method, and investigate also those instincts of the human mind by which man is led to work as if the approval of a Higher Being were the aim of life." (JEVONS, Professor W. STANLEY, of University College, London: *The Principles of Science*, a Treatise on Logic and Scientific Method, pp. 469, 470. London, 1874.)

Here speaks no theologian, no partisan, not even an anti-evolutionist, although Jevons is an anti-materialistic evolutionist, as every man of sense ought to be. Sneers about the carpenter theory, from one who thinks the dicer's the better, are quite out of place, face to face with that majestic peroration of Jevons. Let us be everywhere mercilessly true to the scientific method. Since man does possess instincts by which he is led to act as if the approval of a Higher Being were the end of life, we are to investigate these instincts at least as searchingly as we do those of the bee, the ant, and the beaver.

1. Instinct is an exhibition of intelligence *in* but not *of* the being to which the instinct belongs.

Your bee builds according to mathematical rule; but do you suppose that all the intelligence it exhibits is in an intellect possessed by that insect? Has it planned, has it thought out geometrical problems, and at last ascertained in what style to construct the honeycomb? None of us believe that. We hold that the bee works by instinct, and the difference between instinct and reason is very broad. Instinct never improves its works, but reason does. The bird builds her nest now as she did before the flood, and the honeycomb is the same to-day as it was in the carcass of the lion when Samson went down to Jordan. Instinct copies itself, and no more. It builds better than it knows. But Somewhat knows how well it builds.

Somewhat knows, did I say? What a contradiction it is to affirm that Somewhat knows! Somewhat does not know any thing. Somewhat is nobody. You all admit with Matthew Arnold that behind Conscience there is a Somewhat, but you ask whether behind the Somewhat there is a Some One. When Matthew Arnold says that an Eternal Power not ourselves loves righteousness, he is introducing surreptitiously the idea of a Some One behind the Somewhat. Some One loves; Some One may fight intelligently for righteousness; but Somewhat never does or can love. The eternal Somewhat who loves righteousness! Self-contradictions pervade the most characteristic phrases of Arnold. He constantly introduces the idea of Some One in his citations of Biblical language and in his own sometimes very happy phrases. They are happy chiefly because they conceal and effectively use under the cloak of rhetoric the very ideas he opposes. The Some One he will not name explicitly; but he constantly uses the idea of Some One implicitly. He asserts the existence of a Somewhat, but he will not admit the existence of a Some One except surreptitiously, using the idea though not confessing its existence. Assuredly, if we are to follow Mill in this examination of the eye, with which I opened our discussion, we must suppose that the idea of the honeycomb exists before the honeycomb, as the idea of the eye goes before the eye. The idea must exist somewhere before the plan of these structures existed. Somewhere there must have been an adequate cause of the adaptation of part to part in the honeycomb.

Almost imperceptible creatures in the sea build in the Indian Ocean a goblet. It is called Neptune's cup. Sometimes it has a height of six feet and a breadth of three. It is erected solely by myriads of polypi, fragile animals shrunk within their holes and only half issuing in order to plunge their microscopically small arms into the waves. (POUCHET, *The Universe*, p. 59.) One of these creatures, struggling to keep its position on some reef, made, perhaps, by the graves of its predecessors, begins to build without any consultation with its swarming mates. They all build, and they fashion little by little the base of the goblet. They then carry up the long slender stem. They have no consultation with each other in their homes under the sea. Each works in a separate cell; each is as much cut off from communication with every other as an inmate of a cell in the wards of Charlestown prison yonder is from his associates. They build the stem to the proper height, and then they begin to widen it. They enlarge it, and commence the construction of the sides of the cup. They build up the sides, leaving a hollow within. Every thing proceeds according to a plan. You have first the pedestal, then the stem, then the widened flange of the goblet, then the hollow within, looking up to heaven. The savage passes, and gazes on Neptune's cup in the Indian Ocean, and is struck with reverence. He says in his secret thought: These creatures cannot speak with each other, but they act on a plan as if they were all in a conspiracy to produce just this Neptune's cup. Is the plan theirs, or does it belong to a Power above them and that acts through them? Your poor savage there on the foaming coast of the tropics looks up to the sky into which the cup gazes, and finds the Author of the form of that Neptune's goblet in a Power not of but in the creatures which build it. It is in them, but not of them, for they have no intellect which can conceive what the goblet is; but in isolation from each other they so build their cells that they produce at last a structure having a plan held in view, not only apparently but in fact, from the very first. Even your foremost French materialists find themselves dazed when they stand where this savage does. One of their opponents, writing lately, affirms that Neptune's cup is the noblest challenge that can be thrown down before the school of materialistic evolution. (POUCHET, *The Universe*, pp. 59, 61.) And yet we have men so filled, not with the depth of the sea of thought, but with its mere froth,—so filled with what even the coral insects might rebuke, disloyalty to instinct, that when they stand before Neptune's cup they see nothing to wonder at. By a combination of separate forces, the bioplasts, isolated from each other, in the living tissues which they produce, build the rose and the violet and all flowers; the pomegranate, the cedar, the oak and the palm and all trees; the eagle, the swan, the thrush, the nightingale, the dove, and all birds; the lion, the leopard, the giraffe, the elephant, and all animals; the human eye, and ear and hand and brain, and all men. It is absolutely necessary that the builders of Neptune's cup should be governed by one dominant idea. Does chemistry explain the origin of their common thought? It is also absolutely necessary that all the bioplasts which weave a living organism should be governed by one idea, and that idea differs with the differences of individual living forms. Does chemistry explain the origin of that co-ordinating thought? Neptune's cup alone strikes us dumb. But what shall we say of the mystic structures built by the bioplasts? There is the cup; it is a fact; and the eye is another Neptune's cup; and the hand another Neptune's cup; and all this universe is another Neptune's cup; and out of such cups, I, for one, drink the glad wine of Theism! [Applause.]

2. The instincts of the bee, the beaver, the migrating bird, are found, when scientifically investigated, to raise no false expectations; they all have their correlates; they are never created to be mocked.

3. From the existence of the profound instincts of

conscience, we must infer that they too, when scientifically interpreted, raise no false expectations.

4. But it is conceded that there are instincts in the human mind by which man is led to work as if the approval of a Higher Being were the aim of life.

5. This instinct implies the existence of its correlate, that is, of God as not merely a Somewhat, but also a Some One.

It is not to be supposed that any scientific line fathoms the depths of the nature of the Some One, or of the Somewhat, revealed in the instincts of conscience. But the quality of an infinity we may know even when we cannot know its *quantity*. Knowledge does not cease to be knowledge by becoming Omniscience. Power does not cease to be power by becoming Omnipotence. Space does not cease to be space by becoming infinite in extent. Time is time, although you stretch it out to the infinities and the eternities. Intellect does not cease to be intellect by becoming infinite. The seat of intellect, — that was Paley's definition of personality. We have no better definition than that. Wherever there is a thinker, we know, therefore, that there exists a person. Ideas flame from all quarters of the universe; plans appear in all the Neptune's cups along the coasts of the upper Indian oceans yonder, in the sounding surf of the constellations where the starry dust of the nebula floats as spray. We find there a plan, and here a plan; and wherever a plan, we find an idea; wherever an idea, a thought; wherever a thought, a thinker; and wherever a thinker, a person; and as if you say all has been evolved, we say of necessity, that all has been produced by an Evolver. [Applause.]

6. It is conceded everywhere, that conscience forebodes punishment, and anticipates reward.

7. Those activities of conscience which forebode punishment, and anticipate reward, imply the existence of God as personal. The sense of obligation and the sense of dependence both imply this. The Divine existence, the freedom of the will, and even immortality, Kant called postulates of the practical reason, that is, presuppositions implied in the activities of conscience.

There are organic and instinctive activities of conscience, by which we forebode punishment, or anticipate reward. Who denies this? Not Nero, when he stabs himself, or causes his servant to hold the sword on which he falls. Not Nero, when he hears groans from the grave of his mother, whom he murdered the other day, at Baiæ. Tacitus says, as I recollect at this moment, that Nero, after he murdered Agrippina, heard *sonitum tubæ planctusque e tumulo*, the sound of a trumpet and groans from her grave. He had had no Christian education. He had not been brought up wrongly, and probably did not feel, as Hume did, that it was necessary to explain his qualms of conscience by a shock he received in his youth. Nero had an education drawn out of the black sky and the blood-soaked sods of old Rome; and yet he anticipated the action of the Furies behind the veil. Who will stand here and affirm that these moral fears which in all ages have expressed themselves in what all religions have taught, as to the Furies and Nemesis and the Avenging Fates, and as to what awaits us in time to come beyond death, are not expressions of an organic and ineradicable instinct in man? If God makes an instinct, there is always something to match it. The instinct of a migrating bird finds a South to match it; an ear, sound to match it; a fin, water to match it. We walk directly out upon this universal organic possession of man, and infer the existence of its correlate. The poor bee throws out its antennæ, and touches things near it; and conscience throws out her antennæ, and touches things behind the veil. Conscience makes cowards of us all, not on account of any thing this side the veil, but of something on the other side. But when conscience makes cowards of us all, is it merely of some arrangement of the molecular atoms in the universe, merely of some shiver of the ultimate particles of this inert stuff that we call matter, merely of a Somewhat, or is it of a Some One, that conscience makes us afraid? [Applause.] I have yet to find a materialistic philosopher who does not admit that this foreboding organic instinct is human. This is the way conscience is made; and I undertake to say that it is not bunglingly and mendaciously made.

8. The good, the great, and the poetic minds of the race, in all ages, have described their highest experiences as involving a consciousness of God as personal.

Let your thoughts run through the vistas of historical precedents. Call up Socrates with his protecting Genius, which always told him what not to do. Call up every great poet that has addressed the Muses; call up every orator that has invoked the aid of the gods; remember Demosthenes there on the Bema, looking abroad on the matchless landscape, the temples, the tombs of the men who fell at Salamis, and yet invoking, above them all, the immortal gods. Remember that no public state assembly was opened at Athens in her best days, unless preceded by prayer. A dripping cloud would disperse an audience in the Pnyx, and this because men thought that the portent indicated that the gods were opposed to their assembling. Votive tablets to Jupiter clothed the naked rocks at the sides of the Bema. Even your Napoleon believes in a protecting genius. Lowell pictures the first man in his naturalness, as God-conquered, with his face to heaven upturned. In our highest moments we instinctively speak of a Some One and not merely of a Somewhat. Richter says that when a child first witnesses a thunderstorm, or when the greatest objects of nature, such as the Alps, the Himalayas, or the ocean, come before the mind for the first time, then is the moment in which to speak of God; for the sublime everywhere awakens the thought, not only of a Somewhat, but of a Some One behind it.

Not a Somewhat merely, but a Some One, walks on Niagara's watery rim. The farther up you ascend the Alps, if your thoughts are awake, the more near you come to anticipated communion, not only with Somewhat, but with Some One higher than the Alps, or than the visible heavens that are to be rolled away. There are in the midnights on the ocean, voices that the waves do not utter. I have paced to and fro on the deck of a steamer, midway between England and America, and remembered that Greenland was on the north, and Africa and the Tropic Islands on the south, in the resounding, seething dark; and my home behind me, and the mother-isle before me. Lying on the deck, and looking into the topgallants, and watching them sway to and fro among the constellations, and listening to the roll of the great deep, I have given myself, I hope, some opportunity to study the voices of nature there; but I assure you that my experience has been like that of every other traveller, in the moments when the sublimities of the sea and the stars have spoken loudest. A Somewhat and a Some One greater than they spoke louder yet. The most audible word uttered in that midnight, in the centre of the Atlantic, was not concerning Africa, or America, or England, or the tumbling icebergs of the North, but of the Some One who holds all the Immensities and the Eternities in his palm, as the small dust of the balance. Was that natural? Was it instinctive? Or was this mood a forced attitude of spirit? I should have thought I was not human, if I had not had a tendency to such a mood. I should have been a stunted growth: I had almost said a lightning-smitten trunk, without the foliage that belongs to the upper faculties, without the sensitiveness that comes from the culture of one's whole nature: if I had not felt behind the Somewhat of the material globe the Some One giving it order. [Applause.]

9. In the deepest experiences of remorse there is a sense in the soul of a disapproval not only by a Somewhat, but also by a Some One.

10. It is a fact of human nature, that total submission of the will to conscience brings into the soul immediately a strange sense of the Divine approval and presence as personal.

Pardon me if I ask you to use the scientific method, gentlemen, in the verification of the most sublime fact of human nature. You turn upon the sky your unarranged telescope at random, and you see nothing. Direct it properly, but fail to arrange its lenses, and every thing visible through the tube is blurred. But arrange the lenses, and bring the telescope exactly upon the star, or upon the rising sun, and the instant there is perfect accord between the line of the axis of the tube and the line of the ray from the star, or the orb of day, that instant, but never before, the image of the star or sun starts up in the chamber of the instrument. Just so I claim it to be the fact of experience, — if you doubt, will you try the scientific method of experiment on this subject? — that whenever we submit utterly, affectionately, irreversibly, to the best we know, that is, to the innermost holiest of conscience, — at that instant, and never before, there flashes through us, with quick, splendid, interior, unexpected illumination, a Power not ourselves. The image of the star, or a representation of the sun, is found within the chambers of the poor, feeble, human instrument. You cannot have that nner witness until you have that exterior and interior conformity to conscience ; but whoever has these will know by the inner light that God is with him in a sense utterly unknown before. The axis of the tube must be turned exactly upon the light before you can have the image. An utterly holy choice brings with it a Presence we dare not name. Turn conscience in total self-surrender, gladly and exactly upon the sun behind the Sun, and it is a fact of science that there will inevitably spring into existence a sun behind the lenses, hot enough to burn up your greed and fraud, hot enough to burn up your doubts and those winged creatures of night, scepticism and unrest, which fly through the twilight and not through the noon. [Applause.]

So much as to conscience is known to be fixed natural law. There are undoubtedly in conscience unexplored remainders, both unknown and unfathomable to science. "Conscience and the consciousness of God," says Julius Müller, "are one." But if behind the uncontroverted facts as to the natural action of the highest of all human organic instincts, there are mysteries, the scientific method, with unwavering finger and lips mute with awe, points out in what direction we are to seek their explanation.

"Careless seems the Omnipresent; history's pages but record
One death-grapple in the darkness, 'twixt old systems and the
    Word;
But the yet veiled rules the future, and behind the dim unknown
Standeth God within the shadow, keeping watch above his own."
            Adapted from LOWELL : *The Present Crisis.*

[Applause.]

VI.

THE FIRST CAUSE AS PERSONAL.

THE EIGHTY-SIXTH LECTURE IN THE BOSTON MONDAY LECTURESHIP, DELIVERED IN TREMONT TEMPLE, NOV. 5.

> Ganz leise spricht ein Gott in unser Brust,
> Ganz leise, ganz vernehmlich zeigt uns an,
> Was zu ergreifen ist und was zu fliehn.
> GOETHE: *Torquato Tasso*, iii.

τὸ μὲν ὀρθὸν νόμος ἐστὶ βασιλικός. — PLATO: *Minos*.

VI.

THE FIRST CAUSE AS PERSONAL.

PRELUDE ON CURRENT EVENTS.

THOMAS PAINE has recently been sold at auction in Boston. [Laughter.] We are reminded anew that in many senses infidelity does not pay. At the dedication of the Paine Hall in Boston, in 1875, the editor of an obscure infidel paper said in a public address, reported in " The Investigator," for Feb. 3, of that year, "I will not conceal the fact that we have had a long and difficult struggle. By the unexpected and most generous bounty of our principal benefactor, James Lick, Esq., of California, together with the donations of sympathizing friends from all parts of the country, we have been enabled to erect this edifice, after about fifty years of incessant toil and struggle." Finding that statement in public print, I cited it, and I have been abused by Horace Seaver for doing so, although the paragraph was taken from his own paper. I suppose he thinks my reading that in public was a violation of privacy, his paper has so small a circulation. [Laughter.] But I now hold in my hands another extract from the same paper, and there is much both in and between its lines worth noticing.

A CARD TO THE DONORS AND FRIENDS OF PAINE MEMORIAL BUILDING.

There was a meeting of the board of trustees of the Paine Memorial Building, Oct. 1, 1877.

The trustees met pursuant to notice, Horace Seaver in the chair, B. F. Underwood secretary.

After a statement of the financial condition of the building by Mr. Mendum, and consideration of the same by the trustees, it was voted: That, whereas the call upon the liberal public for contributions to save the Paine Memorial Building, of date June 18, 1877, has failed to elicit any thing like a sufficient sum to meet even the immediate expenses of the building, and seeing no prospect of success in the future, and unwilling to solicit further donations for the building when there seems to be no way to hold it with the contributions we are likely to obtain, therefore we consider it advisable for the interest of all parties concerned, that the building be sold by the mortgagee. This was moved by B. F. Underwood, seconded by Thomas Robinson, and was unanimously passed by the board.

Moved by Mr. Robinson, and seconded by Mr. Mendum, that whereas we have recommended the sale of the Paine Memorial building under foreclosure of mortgage, we decide to revoke all calls for further contributions, and to notify the liberal public that no scrip will be issued by the trustees as a means of obtaining a loan. Passed unanimously.

Moved by B. F. Underwood, and seconded by Thomas Robinson, that if any Liberal shall bid in the building, to be retained for Liberal purposes, we will regard such action as deserving the thanks of the Liberal public; and any effort to obtain contributions or loans by issuing scrip on his personal responsibility would, in our opinion, be worthy of encouragement. Passed unanimously.

The trustees have given much time and attention to the

interest of the Paine Memorial, and made every reasonable effort to obtain money for the building. But the amount received by contributions since we have had control of the building has been small, considering the money needed to pay taxes and interest and meet the necessary expenses. *We have been able to hold the property up to the present time only because Mr. Mendum has generously seen fit to advance the money for the taxes and interest, and thus has postponed the sale of the building.*

The course which the trustees now advise — they can only advise, owing to the heavy indebtedness which puts the building virtually in the hands and subject to the control of the mortgagees — is simply a necessity. Further efforts to hold the property are useless, and we are unwilling to take contributions for the building when we see clearly that even if we were able to meet the present demands, there would be no prospect of preventing its sale at a later date.

                HORACE SEAVER,
                JOSIAH P. MENDUM,
                B. F. UNDERWOOD, } *Trustees.*
                OSMORE JENKINS,
                THOMAS ROBINSON,

BOSTON, Oct. 1, 1877.

                                            (*Investigator.*)

Such is the official statement of the last most painful news concerning this Paine Memorial Hall. [Loud laughter.] I call attention to this ripple on the surface of Boston affairs, not for the sake of this city, where all the facts are well understood, but for the sake of some critics of Boston at a distance who suppose that free thought here has really no place in which it can be wholly without fetters except yonder in the hall just sold by auction. We know better; but it is presumed sometimes in New York, often in Chicago, that the Paine Memorial Hall represented a deep undercurrent here. Now, if it did, why was it not saved, as the only monument to the memory of — well, what shall we say? A crackling pamphleteer who did much for liberty, and who would have been remembered with a degree of honor if his door one night, in a prison at Paris, had not been turned with its back to the wall, and a chalk-mark that indicated his destination for the guillotine been thus concealed.

Had Thomas Paine died in the middle of his career, had he lost his life when death was appointed for him in Paris, undoubtedly we might have remembered him with something of the feeling with which Washington and Jefferson and other leaders of our revolutionary era at one time regarded him. But he lived long enough to show the fruits of his own principles, and to lose the larger part of his earlier friends. Recent discussion has turned a flood of light upon his last years. New York, in Paine's day, had in it men enough willing to conceal his faults — friends of Paine; friends not only of his political but of his religious principles; and who would not have put on record contemporary evidence against him, had not the facts been notorious.

We are not to spend more than ten minutes on this noxious theme, and yet the biographical fact should be remembered that Paine had, in his last years, habits absolutely unreportable before a mixed audience. He was personally filthy, and was at times recommended to bathe as a means of preparing him for company. On one occasion he was hired to soak himself three hours in a hot bath, and he insisted that he did not need the ablutions, when everybody that had called lately upon him had gone away shocked simply by the man's uncleanliness of person. He was a drunkard. He was intemperate not only in the manner common in that day, but roughly, deeply, bestially so. That all this came from his infidel principles, I do not assert, for some men have been drunkards who were not infidels. But Paine, up to the last, continued to be blasphemous toward Christianity. He was proud of his infidelity. I do not suppose that he ever really recanted. It is true that in the last weeks of his life he was constantly calling out, "O Lord, save me!" "O Christ, have pity on me!" He could not bear to be left alone. Even in the high noon, he would shout so as to alarm the house, if left without some one near him. There is evidence that his infidelity sowed the seeds of his bad habits, just as the infidelity of Aaron Burr sowed the seeds of his habits. In Princeton, not long ago, I stood in a celebrated cemetery in an autumnal cyclone, and listened to the whistling of the wind over the grave of Jonathan Edwards and that of Aaron Burr. Who can say that the career of Burr was not the natural outcome of his principles — a systematic course of villany? and who can say that Edwards's career was not a natural outcome of his principles — a systematic course of virtues? I can understand that a man may be born with a dip of the needle that leads him astray among the storms of passion. I have sympathy for those who are wrecked because of deep congenital difficulties. Aaron Burr had these, and Thomas Paine had the same; but I presume neither of them had more terrific passions than Jonathan Edwards or Franklin, and yet in the one case we have lives glorious, and in the other lives infamous.

Among the throng of unimpeachable witnesses of Paine's bestial condition in his last years, is the quiet, candid Quaker, Stephen Grellet, whose life was published in Philadelphia in 1860, and republished in London in 1861. He lived neighbor to Paine; and out of his journal, written in 1809, the very year Paine died, let me read you one extract. I might multiply citations by scores, but this is the most strategic passage in all that has been said: —

I may not omit recording here the death of Thomas Paine. A few days previous to my leaving home on my last religious visit, on hearing that he was ill and in a very destitute condition, I went to see him, and found him in a wretched state; for he had been so neglected and forsaken by his pretended friends that the common attentions to a sick man had been withheld from him. The skin of his body was in some places worn off, which greatly increased his sufferings. A nurse was provided for him, and some needful comforts were supplied. He was mostly in a state of stupor, but something that had passed between us had made such an impression upon him that, some time after my departure, he sent for me, and on being told that I was gone from home he sent for another Friend. This induced a valuable young Friend (Mary Roscoe), who had resided in my family, and continued at Greenwich during a part of my absence, frequently to go and take him some little refreshment suitable for an invalid, furnished by a neighbor. Once when she was there, three of his deistical associates came to the door, and, in a loud, unfeeling manner, said, "Tom Paine, it is said you are turning Christian;

but we hope you will die as you have lived," and then went away. On which, turning to Mary Roscoe, he said, "You see what miserable comforters they are." Once he asked her if she had ever read any of his writings, and on being told she had read but very little of them, he inquired what she thought of them, adding, "From such a one as you, I expect a correct answer." She told him that when very young his "Age of Reason" was put into her hands, but that the more she read in it the more dark and distressed she felt, and she threw the book into the fire. "I wish all had done as you," he replied; "for if the Devil ever had any agency in any work, he has had it in my writing that book." When going to carry him some refreshments she repeatedly heard him uttering the language, "O Lord," "Lord God,' or "Lord Jesus, have mercy upon me!"

God grant that mercy was shown him! Let us show him mercy by remembering his patriotism, and forgetting his anti-Christianity, of no consequence now among scholars, and surely something that ought not to be of any consequence among the ten thousand half-educated young people and operatives who buy the paper-covered "Age of Reason," even yet, as if it were the best book on the infidel side. Not far from Boston a man with gray hairs rose in a meeting where I was the other day, and said that he had burned his Thomas Paine's works and his Voltaire's Philosophical Dictionary, and that he had obtained more light from them in that way than in any other. [Laughter.]

THE LECTURE.

Charles Sumner — *magnum atque venerabile nomen* — in a biography, which, if completed as well as it has been begun, will daze Trevelyan's Macaulay, is represented as standing one morning on the Alpine verge of Italy. He was passing toward the highest glaciers, and noticed at the edge of the way a column, on one side of which were the words *Regno Lombardi*, and on the other *Tyrolese Austria*. He passed the monument, and, suddenly recollecting that he was leaving Italy, rushed backward, and, with the enthusiasm which afterwards sent him into the conflict with slavery, he removed his hat, waved it toward Lago Maggiore and Lago di Como, and toward Rome and Naples, Cicero, Sallust, Tacitus, and all the rest, and said, "I salute thee, Italy," and so parted from the land of flowers. A German, learned, pragmatic, far-seeing, noticing Sumner's action, walked back to the same barrier, removed his hat, and turned his face toward the Fatherland, and said, "Et moi, je salue l'Allemagne." "For me, I salute Germany." (PIERCE, EDWARD L., *Memoir and Letters of Charles Sumner*, vol. ii. p. 125.) Thus opposed in sentiment, these travellers went on. I suppose the German learned to love Italy, if he allowed himself to be bathed at all in Sumner's enthusiasms. It is certain that Sumner learned to love Germany; for beyond the eternal, deadly glaciers, he found a land of cathedrals, stately universities, great religious historic memories, and of patriotism so intense that old Rome never conquered the German forests, but was sent back daunted by Hermann. Our fathers never yielded to the Roman Empire. In Germany Sumner at last, when looking toward Italy from the north side of the Alps, remembered that one meridian joins Rome and Berlin, the North and the South, and that there is no leaving that meridian until we can outswim the bounds of the sky itself. Italy, Germany, are parts of one world. They are fragments of men, they are travellers of a narrow range, they are provincial hearts and intellects, who cannot embrace at once both the cathedrals of the Po and the Tiber and those of the Rhine and Elbe. [Applause.]

Conscience is Italy: reason is Germany; and between them Herbert Spencer and Mansel and philosophers of their school have in every age thrown up Alps, obstructing the natural transition of travellers from one to the other. Conscience teaches that God is a person. The organic instincts of the soul all point to a Being possessing personality, and on whom we are dependent, and to whom we owe obligation. But it is said that reason, strictly interrogated, will not permit us to assert that God is a person; that an Infinite Person is a contradiction in terms; that we cannot call God a person without limiting him; and that to limit him is to deny his infinity and absoluteness.

Many a man in the Italy of conscience has paused at its boundary line on the glacial Alpine heights of thought; and has saluted, as did Sumner, the South, or the moral emotions and instincts; and then turned, with a shiver taking hold of the bones themselves, towards the avalanches of the North, or the icy syllogisms of reason and exact research. If we could only live on the Po always; if we could be effeminate forever; if the South were the only quarter of our nature fit to be trusted; if there were no majestic Northern tribes in the soul that will have reason for their King, — we possibly might be allowed in peace to hold the sentimental and the effeminate faith that God is a Person, and that our hearts and his heart may come into contact, finite with infinite! But a German stands here too, with our Sumner, and he removes his hat, and his salutation is in the opposite direction, and we must move on. It is asserted that hundreds and thousands of armies have tried to cross these Alps, and have perished in the attempt. Herbert Spencer has taken up his abode on the summits, and insists that the avalanches are impassable. Mansel points us to army after army that has been stranded in these snows. Harvard University yonder has one brilliant Spencerian in it, who sits on the Alpine glaciers, and denies that God can be known as a Person, and pities any who seek to find Germany, with its cathedrals and universities and majestic memories, beyond the glaciers. (FISKE's *Cosmic Philosophy*, vol. ii. pp. 395, 405, 407, 409.) His voice, however, is but the echo of Spencer's, although occasionally more articulate than that of the master. It is to Spencer that we must look chiefly, and to Matthew Arnold and to Mansel and to Alexander Bain, for our discouragements as we attempt to cross the Alps of Nescience. I have a faith, and I have it in the name of the general law of the survival of the fittest; in the name of what has been the steady outcome of philosophy, age after age; in the name of the sky of self-evident truths, which has in all its parts but one curve, — that we can cross those Alps. I have four tests of certainty: intuition, instinct, experiment in the large range, and syllogism. By instinct I feel authorized to say that God is a Person. By experiment in the large range I feel authorized to say so. That belief works well. By syllogism, if John Stuart Mill is authority in logic, I am authorized to say that there is a Person, whether he is infinite or not. A God exists who is a Person; and whether we can call him literally infinite or absolute, Mill does not determine; but there is a Person behind the thought exhibited in the universe. Syllogism, experiment, and instinct, three parts of the curve, are thus visible. But I never saw a curve yet that did not run through its fourth quadrant according to the law of its three other quadrants. If we, in discussing the organic instincts of conscience, and in looking into the uncontroverted facts concerning the moral faculty, find a sense of obligation and dependence pointing to a personal God; if all these agnostics, these Spencers, these followers of Arnold, these doubters, some of them orthodox with Mansel, are right in admitting, as they all do, that our organic instincts force us to act as if we were responsible to a Higher Person, — then assuredly we are right in saying that the arc of instinct, in this circle of tests of truth, points to God as a Person. Having a clear view of this one quadrant only, I will dare to project the majestic curve; and into the avalanches, into the mists of the gnarled heights, into all that is Alpine here, I will pass boldly on the line of that quadrant, sure that beyond the summit I shall find a Germany, one with Italy in the beloved South. [Applause.]

1. While it is admitted by the highest authorities that conscience teaches that God is a Person, it is affirmed by a few of these authorities that reason teaches that he is not.

2. It is affirmed that to call God a Person is to limit his infinity; and that an Infinite Personality is a contradiction in terms.

3. In this state of the discussions concerning conscience, if its organic instincts as to its obligations to God as a Person are to be justified intellectually, it becomes of the utmost importance to show that reason as well as conscience teaches that God is a Person.

4. For the purposes of such proof it is highly advisable now to separate the whole topic of Theism into three parts: namely, the proof that the Cause of the universe possesses intelligence; the proof that it possesses unity; and the proof that it possesses infinity.

The question at the outset is not whether God is infinite or finite, but whether he is intelligent or not. It is my object to establish the proposition that conscience reveals not merely a Somewhat, but a Some One; and, having proved from the point of view of instinct that it does, I must now justify the proof by showing that reason can make no objections to that conclusion.

While we are considering intelligence as cause, I leave out of view entirely the inquiry as to its infinity. The question is not even raised, in the opening of an argument such as I am presenting to you, whether God is infinite or not. Can we prove that he is Some One? That is the initial inquiry. Can we demonstrate that there exists in the universe an intelligence not ourselves? After demonstrating that the Cause which stands before the present universe has intelligence, we must ask whether it has unity. After having proved the intelligence and the unity, we must treat the infinity as a wholly different thing. Separate proofs are adapted to these several traits. Do not overload the definition of God when you begin your argument from reason for his existence as a Person.

5. The universe exhibits thought. There cannot be thought without a thinker. The cause of the universe, therefore, is a thinker. And a thinker is a person.

6. But the universe exhibits, so far as human observation extends, perfect unity of thought. Gravitation is the same everywhere, and so are light, heat, and the other natural forces.

7. The universe, therefore, exhibits one thought — and but one.

8. Its cause, therefore, is One Thinker, and but One; that is, One Personal Intelligence, and but One.

The philosophy dominant at Yale College and at Harvard, at Berlin and Halle, at Edinburgh and Oxford and Cambridge, is well represented by these incisive sentences from the ablest book on metaphysics Yale College has given to the world. "The universe," says President Porter, "is a *thought*, as well as a *thing*. As fraught with design it reveals thought as well as force. The thought includes the origination of the forces and their laws, as well as the combination and use of them. These thoughts must include the whole universe: it follows, then, that the universe is controlled by a single thought, or the thought of an individual thinker." (*The Human Intellect*, p. 661.)

Let us pause, and cast ourselves abroad on the wing of imagination, through at least some small portion of the range of truth disclosed by the facts that thought implies a thinker, and that the thought of the universe is one. Take in your hand the mystic instrument called the spectroscope, and bring down light from the two planets which last evening I saw near each other in the infinite azure. Here arrives a far-travelled ray from Mars; here one from Saturn; here one from Sirius; here one from the North Star. It left that orb fifty years ago, and has not paused, and is here at last. Certain metals, when burned, always produce definite dark lines in the colored lights of the spectroscope. We know that zinc produces a line in a particular place, lead in another place, iron in another place; and we bring down this light of Mars, of Saturn, and of the North Star, and here are 'the very lines of zinc and iron and lead. Matter yonder, fifty years distant for light, we thus know to be much what it is here. Meteors have fallen on this earth; the dust of meteors has been absorbed into plants; and, for aught I know, there are in your arm particles that came from Sirius. The universe has light in it; and the laws of light are the same here and at the farthest point visible to the telescope. Light moves in straight lines here and in straight lines there. Gravitation is the same thing here and yonder. We cannot imagine a spot in the universe where the whole is less than a part, or where two straight lines can enclose a space, or where any self-evident truth is false. Thus we feel that the universe exhibits not only a plan, but a uniform plan; it exhibits not only thought, but harmonious thought. It is a thing, but it is a thought; and it is not merely a thought, without further definition: it is one thought, interiorly self-consistent, and not a fagot of self-contradictions. This immeasurable but incontrovertible unity is before our eyes. It demonstrates unity in the thought of the universe, and therefore unity in the Thinker. The universe exhibits one thought, and but one. Its cause, therefore, is one Thinker, and but one; one Personal Intelligence, and but one.

Adhere, without a particle of wavering, to the proposition that there cannot be a thought without a thinker. That is Descartes' fundamental axiom, the corner-stone on which he placed himself face to face with all scepticism and unrest. This is the one point of philosophy where certainty is firmest up to this hour. There cannot be thought without a person. I think: therefore I am a person. There is thought not our own in the universe: therefore there is a Person in the universe not ourselves! The thought is one: the Thinker therefore is One! Sometimes, when I stand under the dome of that truth, I am moved as the constellations never stir me. The old songs once sung in the Temple yonder on a hill that has influenced the ages more than Athens or Rome, come into my thoughts; but even their melodies do not always express fully the enthusiasm which bursts up face to face with the scientific method in our day. We must expand David's outlook upon the universe. No doubt he beheld the moral law more vividly than we do; no doubt he had interior insight such as belongs to that strange race of which he was a representative. The Greek knew art better than we do; compared with him we are uncouth. In contrast with the Hebrew in his best estate, we are morally imperceptive. But these grandeurs of law which God seems to have revealed to us, the Aryan race: these grandeurs of co-ordination which make us, in our fragmentariness of endowment, sometimes almost content with a mere Cosmic Deity, without much thought of a person,— we must unite them all, the modern with the Greek and Hebrew organ-pipes! But the music proceeding from them all together, falling, expanding, filling the dome of the universe — that is but the sound of a shepherd's pipe compared with the melodies that will rise in all full-orbed souls whenever the ages have been taught to look aloft, with adequate intentness, into the azure represented by the simple certainty that there cannot be in the universe thought not our own, without a Person not ourselves; and that, as the thought is one, so that Personality is One. [Applause.] Let us be glad! Let us lift up our hearts! Let us say to the eternal gates of science, "Lift up your heads, that the King of Glory may come in." [Applause.] The day is coming when another age will say this to the gates that have foundations. The day is coming when our transitory stage of thought — simply a sophomoric year in human investigation, and in which we can ask more questions than we can answer — will be looked back upon with disdain. The day is coming when the iron lips of science will utter the words of the Psalmist and the words of all natural law: "Lift up the Gates on which the Pleiades are but ornaments! Lift up the Gates on which all physical immensities and infinities and eternities are but so much filagree! Lift up these Gates, and the King, Immortal, Eternal, Invisible, not ourselves, and who loves Truth, Beauty, and Righteousness, will come in!" [Applause.]

9. The Infinite and the Absolute are words which mean nothing unless we understand by them that which is absolute or infinite in some given attribute.

Stuart Mill was no partisan on the side of Theism, but his dissatisfaction with Mansel's and Spencer's use of the words Infinite and Absolute is well known. Space we call infinite, and we mean not vaguely that it is the Infinite or the Absolute, but that it is infinite in one particular quality, namely, extension. If you speak of space as the Infinite or the Absolute, without stating in what quality the object meant is infinite or absolute, you at once confuse men, because you are not expressing a definite idea. Herbert Spencer, Mansel, and their followers, are constantly telling us we must think thus and so concerning the Infinite and the Absolute. Now substitute for these terms the Infinite Being, the Absolute Being, and very often their expressions will not make sense, or make nothing short of blasphemy. The Absolute, it is said, must contain every thing. "There is a contradiction," says Mansel, "in conceiving the Infinite and Absolute as personal; and there is a contradiction in conceiving it as impersonal. It cannot, without contradiction, be represented as active; nor, without equal contradiction, be represented as inactive." (*Limits of Religious Thought*, Lect. II.) "To define God," said Spinoza, "is to deny him." If we limit God by saying that he cannot do evil, we are putting a bound upon his nature, and he is no longer infinite. Well, all this dense and often deadly vapor arose from a false definition of the Absolute and the Infinite. Say an infinite being, one who is infinite in goodness, cannot be evil, and then say that such an affirmation implies limitation of God! Say that two straight lines cannot enclose a space, and then affirm that such an affirmation involves limitation of the qualities of the object that is infinite, and you confuse all thought, simply because you are yourself confused. The Absolute, the Infinite, are words that have no real significance unless taken in connection with some quality. You must come down to the concrete always to get the meaning of these abstract terms; and the men who sit among the glaciers of the Alps, and tell us the Alps cannot be passed, are sitting, not on the concrete rock, not even on the snow, but on the fog. [Applause.] We speak of time as infinite, but we mean only that it is infinite in one respect, duration. In a similar sense, the one Thinker who stands behind the one thought of the Universe has been termed infinite in the sense of possessing infinite power, and absolute in the sense of absolute, finished, completed goodness and knowledge.

10. It is certain that infinite space is space; infinite time is time; infinite power is power; infinite knowledge is knowledge; and infinite goodness is goodness.

11. What is affirmed, therefore, in calling the Divine Attributes of power, knowledge, and goodness infinite, is intelligible, and involves no self-contradiction.

12. Except the element of infinity, any given quality is the same in its infinite as in its finite development. We cannot adequately conceive the quantity, but we may the quality, of an infinity.

Space is just the same in its infinite as in its finite development. Power is just the same in its infinite as in its finite development. Indeed, we never hear objection to likening God to man brought against this attribute of power. We are told that we are constantly falling into anthropomorphism, but that the tendency of science is to deanthropomorphization. This is getting to be a very popular word, my friends, and so we must accustom ourselves to it. Anthropomorphization, — that means simply an excessive tendency to liken God to man, and deanthropomorphization means the opposite. Spencer and his school often forget that there is anthropomorphism in their own characterization of the Cause of the Universe as a Power. Goethe said we never know how anthropomorphic we are; and I think Matthew Arnold himself does not know how anthropomorphic he is. He is constantly employing phraseology that implies personality in God. "The Eternal not ourselves loves;" "the Eternal not ourselves hates." "The Eternal not ourselves," he personifies constantly. Of course he explains that by personification he means only poetry. But this poetry is organic, instinctive, constitutional. Matthew Arnold's famous proposition, that the Jews did not believe in a God except poetically, that they always knew that there was no Person behind the Eternal Power, not themselves, which they thought made for righteousness, is one of the absurdest of all the eccentricities of the school of Nescience. It really has made no impression on scholarly thought, much as we revere Matthew Arnold and his father. If the father were alive, I think some logical chastisement, at least, would be applied to the son. For his father had a stalwart grasp upon philosophy as well as the historic sense. Dr. Dale told me the other day that Matthew Arnold once said to him in a parlor in London, "I stand about where my father did;" and he considered that remark of Arnold's an indication of a lack of careful habits of discrimination. Dr. Dale replied, "Matthew Arnold, your father believed in the personality of God, and was inspired by that truth to heroic life; and he believed that God has manifested himself in human history; and these things make a difference between your own views and his." And Matthew Arnold's only reply was given in a dazed, uncertain way, "Well, perhaps they do." When Arnold's best expressions agree with the Biblical language, they do so because his instinct moves him toward the attitude which the Bible words express; and that attitude is adoration before God as a Person. That the Jew did not believe God to be a Person, is a proposition just as rational as that the Greek did not believe art to be a worthy field for human effort. We might as well say that the Roman Empire never existed as to say that the Jew did not believe in a personal God.

13. What is inconsistent with goodness will be inconsistent with infinite goodness.

Just here I must pause to show you the 'stalwart manliness of John Stuart Mill. Mansel, believing in Sir William Hamilton's phrases about the Infinite and the Absolute, a few passages which the master never expanded into a system, undertook to assert that God may be so different from man that if there is objectionable truth in revelation we must not apply to it very sternly the human standards of morality. I revere Mansel; but his book on the Limits of Religious Thought seems to me, as it seemed to John Stuart Mill, one of the most mischievous of modern productions. In the name of the limitation of the human faculties and the relativity of all knowledge, — a truth which I do not deny, in the sense in which Sir William Hamilton admitted it, — Mansel affirmed that we never can know intellectually that God is a Person; his goodness may not have laws represented by the self-evident truths of conscience; and, therefore, if difficulties arise in revelation, we must regard the universe as a scheme imperfectly comprehended, and, in case of the Bible, treat it leniently in detail after its general authority is once proved. Stuart Mill, remembering that infinite goodness is goodness, and that what is inconsistent with goodness must be inconsistent with infinite goodness, sat down one day, and wrote his opinion of Mansel's book: " To say that God's goodness may be different in kind from man's goodness, what is it but saying, with a slight change of phraseology, that God may possibly not be good? To assert in words what we do not think in meaning, is as suitable a definition as can be given of a moral falsehood. If, instead of the glad tidings that there exists a Being in whom all the excellences which the highest human mind can conceive exist in a degree inconceivable to us, I am informed that the world is ruled by a Being whose attributes are infinite, but what they are we cannot learn, nor what are the principles of his government, except that the highest human morality which we are capable of conceiving does not sanction them, convince me of it, and I will bear my fate as I may. But when I am told that I must believe this, and at the same time call this Being by the names which express and affirm the highest human morality, I say in plain terms that I will not. Whatever power such a Being may have over me, there is one thing which he shall not do: he shall not compel me to worship him. I will call no being good who is not what I mean when I apply that epithet to my fellow-creatures; and, if such a being can sentence me to hell for not so calling him, to hell I will go." (MILL, JOHN STUART, *Examination of Sir William Hamilton's Philosophy*, vol. i. chap. vii.)

There was an earthquake rent, into which this whole philosophy of Nescience will ultimately be cast in the name of logic, and with the acclamations of all thinking men.

14. The attributes of knowledge, power, and goodness, each of them in an infinite degree, can be intelligibly and without self-contradiction attributed to one Thinker, and to but One; and that One He whose thought the origination and preservation of the universe exhibit.

15. Immense distinctions exist between the Absolute defined as the unrelated, or that which exists out of all relations, and the Absolute defined as the independent, or that which exists out of *one set* of relations, that is, out of all relations of dependence.

16. It is in the latter sense only that scientific Theism asserts that the One Person whose existence is proved by the one thought of the universe, is absolute.

17. Great distinctions exist between the Absolute defined as that which is capable of existing out of relation to any thing else, and defined as that which is incapable of existing in relation to any thing else.

18. It is in the former sense that scientific Theism calls God absolute.

19. It is in the latter that Herbert Spencer, Mansel, and others who deny that we can prove intellectually that God is a Person, call God absolute.

20. This false definition overlooks the distinction between infinite and all, and leads Mansel to Hegel's conclusion that God's nature embraces every thing, evil included.

21. The definition which Mansel and Spencer hold is repudiated by scientific Theism. (See MARTINEAU, *Philosophical Essays, Science, Nescience, and Faith;* President PORTER, *The Human Intellect*, last chapter; President MCCOSH, *The Divine Government;* HODGE, *Systematic Theology*, vol. i. pp. 381–432; NITSCH, ROTHE, TRENDELEUBURG, DORNER, ULRICI, and JULIUS MÜLLER, *passim;* and, especially, MILL's *Examination of Hamilton's Philosophy*, vol. i. chaps. i. to vii.)

22. With that repudiation all the alleged difficulties that arise from asserting the personality of God vanish.

23. Herbert Spencer and his school admit that the Eternal Power, not ourselves, which makes for righteousness in the universe, is omnipresent, self-existent, omnipotent, and in this sense infinite and absolute.

In a recent volume of most searching applications of the scientific method to philosophical thought, Thomas Hill writes: "Spencer says that our belief in an Omnipresent Eternal Cause of the Universe has a higher warrant than any other belief, that is, that the existence of such a Cause is the most certain of all certainties; but asserts that we can assign to it no attributes whatever, and that it is absolutely unknown and unknowable. Yet, in his very statement of its existence, he assigns to the Ultimate Cause four attributes: Being, Causal Energy, Omnipresence, and Eternity. And afterwards he implicitly assigns to it two other attributes — repeatedly expressing his faith that the Cosmos is obedient to law, and that this law is of beneficent result, which is an implicit ascription of wisdom and love to the ultimate cause. All thinkers concede that human reason is competent to discover the existence of an Ultimate Cause, to form the inductions of its Being, its Causal Energy or Power, its Omnipresence and Eternity." (HILL, THOMAS, ex-President of Harvard University: *The Natural Sources of Theology*, pp. 33, 42.)

24. The intelligence, the unity, and in a correct sense the infinity, of the Cause of the Universe, are therefore proved in entire harmony with the scientific method on the one hand, and Christian Theism on the other.

Our best conclusion is adoring silence before the slowly lifting Gates through which the Eternal, who holds infinities and eternities in his hands as the small dust in the balance, is passing into science, into politics, into the perishing and dangerous populations of the world, into the Norse American as well as into the Puritan American, into literature, into woman's heart, into conscience, into the future, and so leading us into that world into which all men haste. He is there, he is here; and our best speech before him, in the name of science, is silence and action. [Applause.]

VII.

IS CONSCIENCE INFALLIBLE?

THE EIGHTY-SEVENTH LECTURE IN THE BOSTON MONDAY LECTURESHIP, DELIVERED IN TREMONT TEMPLE, NOV. 12.

Every man brings such a degree of this light into the world with him, that, though it cannot bring him to heaven, yet it will carry him so far that if he follows it faithfully he shall meet with another light which shall carry him quite through. — SOUTH.

Alle Form sie kommt von oben. — GOETHE.

VII.

IS CONSCIENCE INFALLIBLE?

PRELUDE ON CURRENT EVENTS.

WHEN the Northern Pacific Railway is finished, America will be one thousand miles nearer China than now. Ships from the Oregon coast pass to Saghalien on a comparatively small circle of the globe, while, from San Francisco by the way of the Sandwich Islands, they sail to Japan over the track of a great circle. It is practically settled that a bridge is to be built by commerce across the Northern Pacific — between what two abutments?

On the one hand we have a largely unoccupied country, giving exceptional honor to free labor; offering to the workingman meat every day for dinner; and providing for him a competence if he is industrious and economical. On the other, we have a land supposed to contain from 450,000,000 to 550,000,000 people, suffocated, and many of them starved. It is only a question of time, whether a bridge built between two such shores will be used. It is only a question of time, whether Chinese immigration is to become an important organizing force on the Pacific coast, and redemptive for China by reflex influences from America.

It seems to be forgotten in the United States, that to-day the Chinese are the great colonizers of the East. The natives of Cambodia, Sumatra, Java, the Philippine Islands, Timor, and Borneo, are fading away before civilization. Europeans cannot cope with the insalubrity of the torrid East-Indian climates. The Chinese alone have proved themselves able to maintain vigorous physical life in these regions. They are entering them by thousands, and in some cases tens of thousands, every year, and that in an ever increasing ratio. They are rapidly colonizing Mantchuria, Mongolia, and Thibet. A stream of emigration has of late set toward Australia, New Zealand, and the Pacific coast of America. (*Pamphlet published by English residents of Shanghai, May 16, 1877.*)

Ah Sin comes to California now hungry. He has a little meat to eat every day. Letters in strange characters go back to the rivers of China, containing the wonderful information, which so surprised Charles Dickens when he first landed in Boston, that workingmen in the United States can have meat to eat three hundred and sixty-five days of the year at dinner. Wandering up and down in the Chinese quarter of San Francisco, undoubtedly we meet the vices of heathendom; and of course there is nothing equal to those in Vienna or Paris.

> "For ways that are dark,
> And tricks that are vain,
> The heathen Chinee is peculiar."

But the Californian is not, the Viennese is not, the Parisian is not! Opium fumes are rising here from the corner of the street; they proceed out of a cellar. But absinthe is used among the soft ladies of Paris, I have heard, and sometimes is not unknown in certain spoiled luxurious circles of the United States. Of course the Chinaman is to blame, and we are not. Nevertheless his old heaven of mythology is a rather better one than ours was. Wendell Phillips says that if you wish to know the real traits of nations you must go back to the time when, in paganism, they constructed mythology, and notice what their heavens were. These Chinamen had a Confucius to teach them; and, although that leader of religious thought did not make any assertions about immortality, he did teach reverence for parents and scholarship. The peace and permanence of the Chinese Empire seem to have depended very largely upon that one trait, cultivated by pagan religion. Carlyle says that most European governments, with their sudden revolutions, might take many a shrewd hint from China. Civil-service reform can look to the region of the great rivers, falling from the Himalayas into the Yellow Sea, for examples of competitive examinations for public office, conducted with far more rigor and general justice than are any other political contests on the globe. We had a mythology in which our fathers were represented as in the next life drinking mead out of the skulls of their enemies; as becoming intoxicated in Valhalla, in order the more vigorously to hew each other to pieces; and as rising after the bloodless conflicts to become whole again, and again to become intoxicated and enter into the pastime of hewing limb from limb. We have barbaric blood in our veins yet, and our temptations from Valhalla mead are not ended. Enough has been said of Chinese opium-eaters, but not enough of the greed of English merchants who forced the Chinese trade into the popular sale of that drug.

We wander up and down the Chinese quarter of San Francisco, and hear strange language from roughs. "I would as soon kill a Chinaman as a dog," says one to another. That threat proceeds, perhaps, from some son of an Emerald Isle, emigrants from which New York City considers her chief blessing!

I am aware that when the elephant plucks down foliage from the oak, it is the foliage that becomes elephant, and not the elephant that becomes foliage. Our foreign emigration will be treated in that way, even if it is Irish; but the elephant has trouble, especially in the ostrich stomach of New York. If you insist that he shall endure the assimilation of tons of Irish foliage, in New York, why are you so ready to insist that he shall make no attempts to assimilate a few sprays from the Chinese oak? Many are eager to pass a law prohibiting all Chinese emigrants from acquiring the right of voting here. It is clear from experience that the Chinaman will not be seen as often drunk as the Irishman; it is clear that he will not be seen drunk as often as the low-paid American laborer. Ah Sin has come into collision with low-paid labor on the Pacific coast, principally because he does not get drunk, lives on rice, and sleeps on a board. His vices have come with him, for a poor part of the population around corrupt Canton has crossed over under the spur of the greed of the great Chinese emigration companies. Undoubtedly the women found in the Chinese quarters are unreportably vicious. They are slaves; they are bought and sold to a bondage altogether more ignominious and awful than the black race ever endured on this continent. You sat still while the village of Antioch was burned to the ground on the 1st of May, 1876, and when the Chinese inhabitants there were warned that they could remain in sight of the ashes of their huts only under the penalty of death. Anti-coolie clubs all over California sent messages to officials at Washington, that if measures were not taken to repress Chinese immigration, a similar fate was in store for Chinatown. How many Chinamen are there? Sixty thousand. How many Chinamen are there in California? Two hundred thousand.

What have they done? They hung over the beetling crags of the Sierra Nevadas, and tunnelled them, when the Southern Pacific Railroad was built; and they will do the same work in the gorges of the Rocky Mountains, when the Northern Pacific is built. They were sent down, mired to the waist in mud, to build levees, when San Francisco was threatened with an inundation, and when no white man would take the position. They have performed most of the manual labor in the construction of the railways which have raised the price of the Californian wheat-lands from one dollar to twenty-five dollars an acre. They have monopolized by fair competition the linen-washing of San Francisco. Ah Sin sometimes smokes opium, no doubt, and gambles; but he is mainly concerned in getting a little meat for dinner, and enough money to enable him to go back and bury his bones in China. (See an eloquent paper on the Chinese in California, read at Syracuse, N.Y., by WILLIAM EDWARDS PARK, Oct. 23.)

How can we reach him? By baiting the Gospel hook with the English alphabet. [Applause.] We want a few schools opened in San Francisco. We want a few men to put Ah Sin in a home when his hut is burned up. Here is a man ready to do that, and he is employed by the American Missionary Society. Is he doing any good? When Antioch was burned, he received some of the refugees into his own house. When Ah Sin's hut was mobbed and razed to the ground the other night in the Chinese quarters, he found him some chambers the next day, and helped him through the pinch. The flaming articles in the city press, against the Chinese, this man sometimes answers, and does it eloquently. He is opening schools wherever he can, in the Chinese quarters, and it is found that his position soothes the waters. He is respected by all the better class in San Francisco; and little by little the Chinese come to believe in him. He ought to open twenty schools. Why does he not? He has twenty Ah Sins whom he might succor. He is a man of enterprise, and looks sagacious. Why are his enterprises languishing? His pockets are empty because you have put little into them.

The mayor, and the aldermen, and the politicians, — all honorable men, no doubt, as Cassius was an honorable man, — take note of Ah Sin, and make a law that any laundry-house which delivers linen by a two-horse wagon shall pay one dollar a month tax, and that every laundry-house that delivers by basket and by hand shall pay ten dollars, — laws like those of Philip II. of Spain, against the Moors. When all these things happen, we need to be reminded of what Du Bois Reymond has told us, that nervous influence travels only seventy feet a second in the body. If the floating island we call a whale is harpooned in the flukes, it is a full second, if the fish is thirty-five feet long, before the message can go to the brain, and a return message be sent to the flukes, commanding them to drop into the sea. So wide is America, so broadly do we roll in strength and size in the ocean of time, that one of our greatest dangers is that distance may make us apathetic to our own wounds. We may be harpooned on the Pacific coast, and never know the fact in Boston. [Applause.] The breadth of our land gives most of us the impression that the Chinese question is a bagatelle. Before the harpooned flukes can be dropped into the sea, Ah Sin is mobbed, and his village burned.

Of course the Chinese do not settle here, and are, in some sense, an excrescence on our population. The truth is, however, that irreversible laws of trade are likely to bring a large influx from the suffocated Chinese Empire to our Pacific Coast. One-quarter of the population of the globe lives in that empire. They are dull men, you say. Well, they invented printing, and gunpowder, and the mariners' compass; they were the first to make these innovations, so scholars say; and silk, and porcelain, and a number of other very fine articles, they learned to use before we did. There is behind them a training to orderliness. If they are treated as well as we treat other foreigners under similar circumstances, if our doors on the Pacific Coast are not all to be barbaric ones, if the Chinaman, while he is peacable and industrious, is to be allowed the fair rights of an American citizen, there will be more emigration. Even if he is abused, there will be emigration. I do not know when, but before another centennial, or before the third, there will be an important Chinese element on the Pacific coast. From this certainty, arises the cry of the roughs and hoodlums of whizzing San Francisco, and their call to the people of the East to crowd out the Chinaman, and to make him submit to taxation without representation. He now pays five million dollars annually to government, and corporations, and land-owners, and has no right to vote. When all kinds of indignities are put upon him, and public sentiment, represented by the religious bodies, is decidedly on his side, it is time for the whale —

very like a whale — to give the order for the dropping of the flukes. This question between Irishmen and Chinamen is important, simply as one phase of the labor problem. Surely Ah Sin, while he is industrious, and spends less for drinks than Hans or Patrick, has an equal right to the protection of the police.

Especially has Ah Sin the right to be sent home with a good opinion of Christian civilization. [Applause.] Thousands of these Chinamen have gone back to the pleasant but overburdened land, which stretches its cities almost in a continuous line from the Yellow Sea to the Himalayas. These returned men are scattered through a population, nine-tenths of whom have never heard the central truths of Christian civilization. My opinion is, that there has never been such a strategic opportunity offered to the American Church as now, so far as the evangelization of China is concerned. We have a large Chinese population here, eager to learn, and eager to earn; and with these two purposes behind every Chinaman who lands here, to earn something, and learn the English language, we can draw him enough aside, to be disgusted with his joss-house, and to go back reporting that Christian civilization is better than Asiatic. When we are asked to vote the Chinaman out of this land, we are to remember that for the spread of the highest civilization through a quarter of the population of the globe, California is the door to China. [Applause.]

THE LECTURE.

There is a celebrated oration by Massillon in which he adjures his hearers, at a certain point, to imagine the doors of the temple in which he was speaking to be closed. He then directs them to look upward, and imagine the roof opening upon the azure, and the last day appearing in the infinite spaces. The judgment is set, and you are alone, and how many here will judge themselves to be among the elect? Massillon was philosophically wise in what you call a strange rhetorical device, for it is certain that only in solitude, only in the hush of the visible presence of death and the judgment, can we understand conscience. Voltaire admired this oration of Massillon's. When Louis XIV. heard it in the chapel at Versailles, he covered his face with his trembling hands. When it was delivered in the Church of St. Eustache, in Paris, the whole audience rose with a sudden movement, uttering a deep, wailing cry of terror and faith, as if a thunderbolt had suddenly fallen in the middle of the temple. (MASSILLON, *Sur le Petit Nombre des Élus.* See Le Cardinal MAURY, *Essai sur l'Eloquence de la Chaire.*)

The inner sky, like the outer, is studied best in its depths, when God shuts up the world in his ebony box, to use George Herbert's phrase. Our secret thoughts are rarely heard except in secret. No man knows what conscience is until he understands what solitude can teach him concerning it. Thomas Paine could not bear to be left alone. Many an inmate of prison-wards dreads solitary confinement more than any thing else. The secret of solitude is that there is no solitude. At Mount Holyoke, at Wellesley, and in Vassar College, every pupil is advised to be a certain period each day alone, with the Bible and with God. If any here think they have sounded the depths of their own natures, if any suppose they have mapped all the constellations in the heavens even of Transcendentalism, let them thoughtfully and persistently try the experiment of looking out of the cool, deep well of solitude into the sky; and even at noonday they will find there vast depths, and constellations visible, fit to blanch the cheeks. These are facts. That is the way human nature acts. Therefore Massillon shall call pause here to-day, while I ask whether conscience is infallible, and whether in its infallibility we have not the touch and the vision of a personal God? Imagine the doors closed, and the judgment set.

1. Conscience is that which perceives and feels rightness and oughtness in moral motives, — that is, in choices and intentions.

2. The word motive has three meanings, — allurement, appetite, intention.

3. When Cæsar crossed the Rubicon, his allurement, or objective natural motive, was the political prize of supreme power in the Roman Empire.

That was wholly outside of himself. He was not responsible for its existence. Nevertheless it was a motive to him, in the sense of allurement.

4. His appetite, or subjective natural motive, was

made up of his constitutional endowments, including ambition and love of power.

He did not create these. They were wholly outside the range of his choice.

5. In neither of these senses of the word motives does conscience judge them; and in neither of these senses are we responsible for them.

6. But Cæsar's intention in crossing the Rubicon was determined by himself: he put forth his own choice; his preferences or moral motives were wholly his own; and were, as he was pleased to make them, either honorable or dishonorable, good or bad.

7. In this sense of the word motives we are responsible for them, and conscience does judge them.

8. Most mischievous confusion of thought arises from not distinguishing the three things signified by the word motives.

Here is a library, and there a whiskey-den or some other Gehenna breathing-hole. I stand in the middle of the street between them, and freely choose into which I will go. I am a human being. There is whiskey yonder; that may be an allurement. I did not put it there; I am not responsible for its intoxicating power. In one sense it may be called a motive to me; but call it simply an *allurement*, and you will speak with greater accuracy. I have disordered appetites; I have inherited bad blood, it may be, from some intemperate ancestor; and I have not taken care of myself; I have allowed nerve-tracks of intemperance to groove themselves into my physical organism, and there is a powerful tendency on the part of my diseased blood toward that place of temptation. I am not responsible for that. I may have been for the fostering of the tendency, or for the undue intensifying of a natural appetite for excitement. But I did not create the constitutional tendencies which move me. If you call these motives, I am not responsible for them; but outward allurements and inward appetites are not the only forces concerned here. Finally, I make up my mind that I will go in there, and drink. It is my *intention* to go in there, and drink. I put forth a choice. I step freely into that place of temptation. I come out a beast. I am responsible for that. I did that from my own intention, and by my own motive, choice, and purpose, in obedience to an elective preference which I put forth. Here is motive, in the sense not of *allurement*, or *appetite*, but in that of *intention;* and this is what conscience judges. Intentions are the zenith of the human inner sky; and looking up into their depths whoever uses the eyes of science will see a Throne, and the books opened, and a judgment bar. These are incontrovertible facts of human nature.

But here is a library, and there are books in it. I know their value. They are a motive to me, in the sense of allurement, or what the writers on ethics call an objective natural motive. But I did not place the books on the shelves; I am not at all responsible for their attractive powers; they are an allurement only. Moreover, I have intellectual curiosity, or some moral desire, it may be, for study; and this moves me toward the library; but I am not to be praised for that. Perhaps I inherit it. I may have intensified the power of these natural desires; but an intellectual and moral equipment belongs to me as a human being; and as a motive I am not responsible for it; and conscience does not judge me for its possession. It is an appetite, or what the books call a subjective natural motive. But now I make up my mind to go into that library. That is my act. I intend to go there, and I have the good motive of obtaining information to increase my usefulness, or, it may be, the base motive of acquiring knowledge to enlarge my powers of self-display. I have a motive, a secret intention, a purpose, which I alone am putting forth, and for which I alone, before conscience, am responsible. Thus, in the whole range of his free intentions, a man finds conscience always standing before him, with the doors closed, and the skies opened, and the judgment set.

You know that these are facts; and, if you please, they are just as important facts as any thing about the Ichthyosaurus or the Plesiosaurus. They are as important as speculations about any object in the Zoölogical Museums in Cambridge yonder; they are as important as any thing we touch with the microscope or scalpel; and, indeed, quite measurelessly more so. Let us distinguish the three classes of motives, or allurements, appetites, and intentions; and be unalterably sure that, however much force the first and second may have, we are responsible for the third.

A distinguished theological teacher once illustrated the difference of the three kinds of motives by the case of a boy climbing an apple-tree to steal apples. The apples are the objective natural motive; the boy's appetite is the subjective natural motive; his intention is his moral motive. The boy climbs the tree to get the apples, and there is his exterior natural motive. He climbs the tree because he is hungry, and there is his interior natural motive. He climbs the tree because he has a mind to, and that is the motive for which he is responsible. [Applause.]

A shallow, and often vulgar, semi-infidel paper in Boston has lately discovered that motives and intentions are not the same, and that we are not responsible for our motives. Certain haughty critics of this lectureship, who assert that we are never responsible for our motives, will do well to look at any common vocabulary of philosophy, such as Flemming's or Krauth's, under the word Motive, and they will find that the distinctions on which I have now insisted are not invented for the occasion, but are as old as Plato.

But so closely does the topic of Conscience touch that of the Will, that we need yet further definitions. We are now on contested ground, where ambiguity of phraseology has been an exhaustless source of debate.

6. Will is the power of putting forth choice, or imperative volition.

10. Choice is agreeable elective preference. It is preceded by a comparison of at least two objects,

and by an excitement of the sensibilities in relation to the objects compared. It may be followed by acts tending to gratify the choice. All choice implies ratherness. Therefore the choice of an object involves the refusal of its opposite.

Choice cannot be defined. You cannot define the word white. You can give a nominal definition of it, but not a real one; and so of choice we can give no real, but only a nominal definition. However, let choice be called agreeable elective preference. It is important to put into the idea of choice this trait of agreeableness, for mere resolution is not choice. The love which the nature of things and the Scriptures command us to have for virtue is choice; that is, we are so to choose it as to be happy in doing so; we are to make duty a delight. We are to choose good, and to be glad in it. No man chooses good unless he likes to choose it. Every choice implies free ratherness. That act of the will which we call elective preference is always agreeable. Forced preference is a phrase involving self-contradiction. Agreeable elective preference, that and nothing less, is choice. This meaning harmonizes well with all the proverbs of the nations. "What a man loves, he is." Show me what a man chooses, and I will show you what he likes most, and what he is most like.

(1) Our sense of what ought to be, invariably requires us to choose what conscience commands.

(2) To choose is to love.

(3) Since, therefore, there is a personal God in conscience, to follow the still small voice is not only to

believe that God is a Spirit, and that he touches us, but to be glad that he is and does so.

These three propositions are the unassailable foundations of the religion of science.

As to the truth that all virtue consists in choice, New England philosophy stands in contrast with European. Very often, by choice, European philosophers mean volition, resolution; and when New England philosophy, represented by Transcendentation as well as by Jonathan Edwards, asserts that all virtue consists in choice, it was once not always understood in Scotland, and still less often in England and in Germany, that by choice Edwards meant agreeable elective preference of virtue. We say that all sin is in choice, when we mean by that word an agreeable elective preference. We *choose* darkness rather than light only when we love it more. We *choose* light rather than darkness only when we love the latter the less. The innermost love of the soul is indicated by its elective agreeable preference.

11. Intention may be defined as a resolved choice. When the fixed plan of executing that choice is entertained by the mind, the intention is called a purpose.

12. *Motives, defined as intentions, choices, and purposes, are perceived by conscience to be right or wrong.*

Accurate observation of our mental and moral experience demonstrates that we have within us a faculty which points out the difference between right and wrong, in our intentions and choices, thus defined, as the faculty of physical taste points out the difference between the sweet and the bitter. We have, therefore, in human nature itself one sure support for a religion that will bear the examination of the ages. I am appealing to proof-texts from the oldest Scriptures, that is, the nature of things. Some silly person wrote the other day from Cambridge, England, that in this lectureship it is not thought worth while to cite the Bible, and that the attempt is merely to build up a religion without any reference to the Scriptures. The castle of the Scriptures stands here, and there are defenders in it. After nineteen centuries of victorious repulsion of assaults, it needs no assistance from me. But haughty Science comes forward, with other weapons; and I have been placed here by friends of free discussion, not to instruct them in any thing Biblical or scientific that they do not know, but to go down into the field before the castle, and with the very weapons of these arrogant foes to meet them in their own redoubts. [Applause.] When religious science, with only the equipment that natural science can give it, comes into the open field, foregoing the aid to be derived from its own fortress, and willing to meet all objections on the ground of bare Reason, it is merely a begging of the entire question to say that the Bible has been given up. On Sundays I go into that fortress, if you please. [Applause.]

It will not now seem other than scientific to assert, in view of the propositions already put before you, that: —

13. All sin or holiness consists, not in volition, but

in elective preference, choices, intentions, moral motives.

External acts possess expediency or inexpediency, harmfulness or mischievousness; and their character in these respects I must ascertain by a combined use of judgment and conscience. I do not know by conscience whether you are a good man, or a bad man; I do not know by conscience whether I ought to defend the President's Southern policy or not. It is a question of judgment, what I ought to do concerning the South. I must gather all the facts; I must look at human experience; I must take the entire light I have, or can get; and then, in the action I choose, conscience will tell me whether my intentions are good or bad; that is, whether I am willing to follow all the illumination I possess or can obtain, or not. I know what my motives are in my political action; I know what I intend to effect; and you all judge men by their intentions in the last resort.

Conscience guarantees only good intentions. Are they enough? If conscience, when truly followed, does not give us soundness of judgment, really it is not a very important faculty, you say. But let us notice what can be proved beyond a doubt; namely, that a man who follows conscience we are able to respect, and that we are not able to respect a man who does not follow it. It is a stern fact that unconscientious intentions no human being is able to respect. We cannot help calling a man respectable who is possessed of good intentions; nor can we help finding him not respectable who is not possessed of them. There is Stonewall Jackson, and here is John Brown. Let us suppose that Stonewall Jackson believes that John Brown is utterly honest; and let us assume that John Brown believes the same of Jackson. Brown's action appears to Jackson to be very mischievous, and Jackson's action appears to Brown to be equally so. In fact, they are crossing bayonets in a civil war; but they are both men of prayer, men of confirmed religious habits, and we have reason to believe that they are endeavoring to be conscientious. I do not believe Stonewall Jackson followed all the light he had; nor do I believe John Brown did. But suppose that Jackson did follow all the light he had, or could get, and suppose that John Brown did, and that each is convinced of this fact as to the other: then, although they are ready in the settlement of practical measures to cross bayonets, you cannot help their coming together when the measures are settled, and shaking hands with each other as respectable men. You know that to be the fact. External acts differ to the degree of crossing bayonets; but, as each does the best he knows how, each respects the other, and absolutely cannot help doing so. This is a singular fact in the soul; but this is the way we are made. We find that Governor Wise, when he looked into the eyes of John Brown, and saw honesty there, and that others who noticed his mood in his last hours, were thrown into a kind of awe by that border warrior. He meant right; and respect for that man's soul is not confined to the circle of the mountains between which he lies in my native county in Northern New York. I have heard the summer wind sighing over the grave of John Brown, and have stood there and gazed upon Mount Marcy and Whiteface and Lake Placid; but because I believed that this man's conscience was a Lake Placid, and his resolution to follow it firm as Marcy, firm as Whiteface, firm as any of those gigantic peaks in my native Switzerland, I felt sure that his soul was marching on [applause], and that when his spirit smote slavery, the tree after that was timber. [Applause.] It did not fall at once, but it was no longer alive.

There was a persecutor of the early Church who verily thought he ought to do many things against Christianity. He himself teaches us that he needed pardon, but that mercy was shown him because of his ignorance. Who will say that he did not suppress light? Not I. He did immense mischief while his judgment was not corrected; and if he suppressed light, or tutored it, his choices were not good. This is most dangerous ground. I know on what treacherous soil I tread, unless definitions are kept in view. Choice means love; conscientiousness is glad self-surrender to a personal God in conscience, or to what ought to be in motives. Let us take the precaution of using pictures as well as metaphysical phrases. There is a point in the bounding, resonant Androscoggin at which is an island, and on it lives a hermit. Twenty savages are sailing down in the midnight to surprise him and put him to death. A Maine legend says that he puts a light below the deadly Lewiston waterfalls that lie just beyond his island. The Indians think the torch is in his hut; row toward it; and all of them make a sudden, dizzy, unexpected plunge to death. The Indians were in one sense right; they wanted to land where the light was; but the light was below the falls, and not above. It is tolerably important to know where the beacon is, whether below or above the cataract.

Rothe well says that the supreme sin is the suppression of light, or the attempt to deceive the cognitive faculty.

Conscience is your magnetic needle. Reason is your chart. But I would rather have a crew willing to follow the indications of the needle, and giving themselves no great trouble as to the chart, than a crew that had ever so good a chart and no needle at all. Which is more important in the high seas of passion, the needle Conscience or the chart Reason? We know it was the discovery of the physical needle that made navigation possible on the physical seas; and loyalty to the spiritual magnetic needle alone makes navigation safe on the spiritual seas. When we find a needle in man through which flow magnetic currents and courses of influence that roll around the whole globe and fill the universe, causing every orb to balance with upright pole, we know there is in the needle something that is in it but not of it; and we may well stand in awe of it, and refuse to tutor it. Show me a crew without a chart, but willing to follow the needle, and I will show you safe navigators; but show me a crew with a chart who will not look at the needle, and I will show you navigators near wreck. Conscience requires every man to mean well, and to do his best. It requires us to follow not only all the light we have, but all we can obtain, and to do so gladly. Give me a Lincoln, and I will trust a nation's welfare to him, for the judgment of the leader will grow right by following all the illumination he possesses. Give me a Lord Bacon, with never so wide windows of merely intellectual illumination, and no purpose of doing the best he knows how, and I dare not trust him where I would trust a Lincoln of far inferior intellectual powers. You know that it is a right heart that, in the end, makes a safe head; and the ancients used to say that the punishment of a knave is that he loses good judgment. [Applause.]

14. John Stuart Mill, although a determined opponent of the intuitional school in philosophy, admits that at least one of our perceptions, namely, that a thing cannot both exist and not exist at the same time and in the same sense, is "primordial," and not the result of experience.

The assumption of the associational school in philosophy is that all axioms are merely the result of experience, and might have been different if we had been boxed about differently in our contact with life. It has been taught that there may be worlds where two and two do not make four, and where the whole is not greater than a part. But John Stuart Mill, who is the foremost Coryphæus in the associational school of metaphysics, admits that our incapacity of conceiving the same thing as existing and not existing "may be *primordial*. All inconceivabilities may be reduced to inseparable association combined with the original inconceivability of a direct contradiction." (MILL, *Examination of Sir William Hamilton's Philosophy*, vol. i. chap. 6.) This is a far-reaching concession. Here is a square; it cannot be a circle. Here is a circle; it cannot be a square. At one and the same time one and the same object cannot be black and white. Mill says this perception is primordial. It does not arise from experience; a thing must exist or not exist; and the proposition that a thing can exist and not exist at the same time and in the same sense, Mill says is perceived to be true by a primordial peculiarity of the mind. If any one of Kant's or Hamilton's unsuccessful critics is dissatisfied with the use of the word intuitive, I will be satisfied with the use of Mill's word, primordial.

15. If we are so made that the distinction between a whole and a part is primordial, or perceived by a power which we possess antecedent to all experience, it may be proved that conscience, within the sphere of motives or intentions, is infallible.

16. To follow conscience is to suppress no light, that is, to follow the whole and not a part of our light.

17. Precisely this primordial or intuitive knowledge, therefore, is that which is involved in the

direct vision conscience has of the moral character of motives.

18. Every man does know infallibly whether he means to do the best he knows how, or not, in any deliberate choice. By a primordial faculty not derived from experience, he knows whether the purpose or intention of following all the light he has exists or does not exist in his mind.

Called upon to choose what I will do, I have a certain amount of light. The interior of my soul is like the interior of this Temple; and now I am to decide whether I will act according to all my illumination candidly or not. I know whether I turn away from the light or not. I know whether I look on the whole or on a part only of this illumination. Mill says that our direct perception of the difference between a whole and a part is primordial. Well, I affirm that if it is primordial in physical things, it is primordial in spiritual things. I have illumination, and I know whether I suppress a part of it. I know whether the whole is taken as my guide, or whether I turn away from some section of the radiance. The distinction between the whole and a part is primordially perceived in the field of mental vision as certainly as it is in the field of physical vision. It is just as infallibly perceived there as here. The perception in both cases is a direct vision of self-evident truth.

There is an ancient Book that speaks of the mischief of the suppression of light. There is a Volume which says that "this is the condemnation, that light is come into the world, and men love darkness rather than light." All this is said in connection with the most subtle doctrines concerning "the Light that lighteth every man that cometh into the world." I find, therefore, that this general view of conscience, as something which always pronounces it right to follow all the radiance we have, and wrong to suppress light, coincides marvellously with the profoundest thought of Christianity, that whoever tutors "the Light that lighteth every man that cometh into the world" is acting against Light which "in the beginning was with God, and was God."

19. Conscience invariably decides that to suppress light is wrong, and that to follow all the light we have or can obtain, and to do so without the slightest tutoring of the radiance, is right.

20. The perception of the difference between meaning right and meaning wrong in this sense is primordial, or intuitive; and the difference exhibits the three traits of all intuitive truth, — self-evidence, necessity, and universality.

If the proposition that a whole is greater than a part is self-evident, necessary, universally believed as soon as men understand the terms, so the distinction between following the whole or a part of our light is self-evident, necessary, and universally admitted as soon as men understand the terms. Therefore, if you use the word primordial as to the small things of physical vision, I will use it as to the great things of spiritual vision. If you use the word necessary as to self-evident truth here, I will use it as to self-evident there. If, in the same connection, you use the word infallible here, I will rise into the upper heaven, and use the word infallible there.

21. With equal clearness conscience always points out that we ought to follow good motives, and not follow bad as here defined.

22. Within the field of intentions or the moral motives, therefore, conscience has the infallibility which belongs to the perception of self-evident truths, and in Kant's language "an erring conscience is a chimera."

There are men who do not know that when they tutor the magnetic needle they are tutoring currents that enswathe the globe and all worlds. There are men who do not know that when they tutor conscience they are tutoring magnetisms which pervade both the universe of souls and its Author. Beware how you put the finger of special pleading on the quivering needle of conscience, and forbid it to go north, south, east, or west; beware of failing to balance it on a hair's point; for whoever tutors that primordial, necessary, universal, infallible perception, tutors a Personal God. [Applause.]

VIII.

CONSCIENCE AS THE FOUNDATION OF THE RELIGION OF SCIENCE.

THE EIGHTY-EIGHTH LECTURE IN THE BOSTON MONDAY LECTURESHIP, DELIVERED IN TREMONT TEMPLE, NOV. 19.

Handle so, dass die Maxime deines Willens jederzeit zugleich als Princip einer allgemeinen Gesetzgebung gelten könne.

KANT: *Prak. Vernunft*, vii.

The idea of a Supreme Being, infinite in power, goodness, and wisdom, whose workmanship we are, and upon whom we depend, and the idea of ourselves, as understanding rational beings, being such as are clear in us, would, I suppose, if duly considered and pursued, afford such foundations of our duty and rules of action as might place morality among the sciences capable of demonstration, wherein, I doubt not, but from self-evident propositions, by necessary consequences as incontestable as those in mathematics, the measures of right and wrong might be made out.

LOCKE: *Human Understanding.*

VIII.

CONSCIENCE AS THE FOUNDATION OF THE RELIGION OF SCIENCE.

PRELUDE ON CURRENT EVENTS.

THE Roman pagan Epictetus wrote: "Dare to look up to God, and say, Deal with me in the future as thou wilt; I am of the same mind as thou art; I am thine; I refuse nothing that pleases thee; lead me where thou wilt; clothe me in any dress thou choosest." (EPICTETUS, book ii., chap. xvi.) Modern civilization is being clothed in a robe of great cities. It ought, if it has the wisdom of Epictetus, to look up and say to Almighty Providence, "Clothe me as thou pleasest; I am of the same mind as thou art." Perishing and dangerous classes are accumulating in cities; and in cities, therefore, the problem of the right management of these classes is to be solved. It appears to be the purpose of Providence, to gather men more and more into cities, and to save them there. City philanthropic and religious effort for the masses of plain and poor men in cities is demanded, and will certainly be honored of God. So far as my knowledge extends, the most important advances that have been made in America in reaching the unchurched masses in large towns, have been effected through the Young Men's Christian Associations and city tabernacles. A luxurious age naturally holds the opinion that the Church should be a place for the select, as well as the elect. But the opinion of Providence concerning modern times appears to be that the telephone and the railway and the telegraph, at their points of intersection, are to draw average men together in suffocated crowds. Already in the United States we have one-fifth of our population in cities, and we had but one twenty-fifth in great towns in 1800.

Five things appear to me to be incontrovertible: —

1. That the American Church, as organized under the voluntary system, is not reaching the unchurched masses in our large cities with due effectiveness. I do not deny that the churches reach the masses; but they are not *so* reaching them as to make the perishing and dangerous populations safe under American suffrage, under our loose government by careless elections, under our elective judiciary, and with the rising importance of the questions between labor and capital.

2. That the unchurched masses, or unseated parishioners in great towns, have often in many cities of Great Britain and the United States been reached effectively when addressed earnestly in tabernacles and in free halls for evangelistic services, by Young Men's Christian Associations, or by the union of churches; and that a large floating population in our

cities is much more likely to be brought within hearing of religious truths in this way than by purchased pews of their own in places of worship.

3. That, if the American churches can reach the unchurched masses of our cities, they ought to do so; and that to neglect an opportunity, growing wider and wider every year, for the management of the perishing and dangerous populations in a Christian way, is a crime. We have opportunity open in one direction. It does not suit us, or not all of us; but it is the instrumentality which has thus far been most successful; and, until some more fruitful method of labor offers itself, Providence seems to indicate that tabernacles have a mission.

4. That when the masses who do not attend the churches have been reached through tabernacles, they are more easily reached through the regular churches.

5. That there ought, therefore, to be no more rivalry between the work of Young Men's Christian Associations and city tabernacles conducted with evangelical and earnest leaders, on the one hand, and the work of the regular churches on the other, than between the fingers and the palm. [Applause.]

It may be that I venture something in defending these propositions; but you will not accuse me of selfish motives, for I have no church and no deacons and no tabernacle. I am looking only to the fact that America needs management in her great cities. If we can manage the one-fifth of her population who live in large towns, we can take care of the rest; but if we cannot manage that perishing and dangerous part of her population, the black angels assuredly will.

What has been done in Boston? Let us answer that question two years hence. Enough time has elapsed in Great Britain to test the work of the American evangelists there. I hold in my hands an opinion of a revered Englishman, who has just been instructing Yale College, Dr. Dale, and which I shall venture to read. It is well known that Mr. Spurgeon, who was at first somewhat shy of indorsing the evangelists' work in Great Britain, now does so most thoroughly. He has followed the good English rule, and under the test of experience the work approves itself to his very experienced judgment. And here is another judgment, also experienced: —

"It is with the liveliest satisfaction and the deepest gratitude," says Dr. Dale, " that I bear witness to the reality and permanency of the impression made on our community during the fortnight of Mr. Moody's stay. Fourteen hundred persons were converted, and united with the churches of Birmingham; six hundred others had received religious impressions, who did not then profess full light and joy in believing. Before Mr. Moody's departure a converts' meeting was held, to which no one was admitted except by ticket. Cards were distributed among the fourteen hundred present, and each new convert was requested to write upon his card his own name and address, and the name of the church with which he desired to connect himself. Of these cards I received a hun-

FOUNDATION OF THE RELIGION OF SCIENCE. 205

dred and twenty. I preached a converts' sermon on Acts i. 15, last clause: 'The number of names together were about a hundred and twenty.' I was unable to visit and examine them personally. I accordingly distributed the cards among several members of the church, and sent them out to examine and report. Many of the letters which I received in response read like romances. Tales of want and woe, and struggle with temptation, and lives of sin! The converts were of various social positions, but the large majority were profane, drunken, irreligious, and even immoral. Fearing lest hesitation and delay might arouse on their part suspicion of my confidence in their sincerity, I received them into the church without the usual probation. Between a hundred and twenty and a hundred and thirty were thus received. I expected numerous defections among these, owing to the class of society to which they belong and the imperfect examination upon which they were admitted. Two years and a half have elapsed. The fruits remain. I hear of profane women, who were the terror of their neighborhoods, living sweet and lovely lives, and of drunkards reformed. I went over the entire list with assistance just before leaving for America, and it resulted from that investigation, that not more than eight, or, at the most, nine, of the entire number, have fallen away. Moreover, the impulse which Mr. Moody's visit gave to our whole church life still continues. It is one of the greatest disappointments of my visit to this country, that I have been unable to meet a man whom I learned, in the brief time he was with us, to love and to esteem." [Applause.]

This city is not cold or haughty, except on the surface. Boston desires safety in her new enterprises, and applies stern tests, indeed, to all religious proceedings. But with the experience of Birmingham, Liverpool, London, Edinburgh, Glasgow, Philadelphia, New York, and Chicago, behind her, ought not Boston to drop a little of her iciness of reserve, and see to it that the fruits of last winter, already reaped, are bound up, and other laborers sent into the harvest, white, at this hour, for the sickle? [Applause.]

Have we visited the five thousand whose names and residences were ascertained and recorded at a meeting of converts? I am not given to counting the results of revivals; but it is very well known that those who have examined the facts most elaborately assert in public prints, over their own signatures, that in the Tabernacle meetings last winter at least five thousand persons made up their minds to do their duty. When, by other methods, not one of which do I underrate, have the churches of Boston done as much? When have you reached the intemperate as well as you did last winter? When, especially, have you exhibited any such blessed activity in personal visitation among the degraded? It is a fact that the most leprous quarters of this city were visited by noble women, and that again and again brands were snatched from the burning. It is the subtle temptation of our luxurious civilization, that we are above such work, and that, because we are above it, we like to have a theology preached which never asserts that a man can be ruined, and especially not that a woman can be. This gospel of luxury, this unscientific liberalism, this tendency to make religion genial, whether it is true to the nature of things or not, is a temptation which cannot be conquered unless we go down face to face with the scientific method to the edges of the Korah's pits where men are swallowed up alive. When the Church has due practical activity, she will have, because she will be obliged to have, a scientific theology, tender as the dew, clear as the sunbeam, serious as the lightning. Christianity once in action can never be content with limp and lavender liberalism; an unaggressive indifference to the fact that men can be ruined; or a religion that believes in plush or velvet, and the genial, rather than in usefulness, and the scientifically true.

Surely the activity of the churches here last winter was sufficient to repay them for all they did. If no good effects had come from it except the quickening of practical Christian work, that alone would have been worth all the effort put forth. What a good thing it was to see all denominations united! Some, from whom we could have expected only silence, were on our side. When the churches are accused of lacking union, let the union efforts made in our cities and in the tabernacles repel the charge.

There is a Lord's table in the Church; and, when invitations to it are given, all denominations, or very nearly all, are brought together. [Applause.] To me, the Church is best represented by the union signified by that common invitation. In the alcoves of a great library, sometimes we have a recess filled with books on Greece; then another with books on Rome; but all the recesses open into one hall. So the different denominations are but recesses in one vast temple; they all open out into one great palace floor, up and down which, in stern times when we really do our duty for the perishing and dangerous, our Lord walks, arm in arm, not with the Baptist, not with the Presbyterian, not with the Methodist, not with the Congregationalist, not with the Episcopalian, but with the whole Church, which is his living garment. [Applause.]

You say that the work done in tabernacles and Young Men's Christian Associations is often superficial. Will you see to it that men are invited into activity in these places who have proper equipments? Some men say the wrong thing technically, in their expression of religious truth, and yet make the right impression; and some men say precisely the right thing,—very martinets of language in theology,—and make a wrong impression. [Applause.] Is it not a matter of amazement, when five thousand persons here in Boston have been brought to a resolution to do their duty, and a great part of them have united with the church, that we should hear from the collegiate city of New Haven very little response, except the statement that Mr. Moody's views are not sound on the matter of the Second Advent? I had known Mr. Moody two years before I knew what his views on the Second Advent were; and, if his great usefulness continues, I shall know him twenty years longer before I care. Provided his devout effort is blessed of God, as it has been; provided he is endowed from on high with the capacity to reach, through his tenderness of heart, through his marvellous practical sagacity, and through the activity that almost made him an invalid here in Boston, working until midnight, and carrying his labor through with a zeal that no man could understand who did not help in it: provided he continues labor of that sort, I, for one, shall consider it an honor to Boston if she can help him a little, and not criticise him at all. [Applause.] He is abundantly able to do without the appreciation of this city, where, after all, he has been appreciated well; and where his work, I think, has been as remarkable as in any other city he ever visited.

Twice the Tabernacle has been open this season, and twice it has been well filled. Hundreds go there who do not go to the regular churches. The unchurched masses are to be criticised for not being willing to go to established places of worship. Every church in America is the result of the voluntary system. We shall have, no doubt, luxurious churches in our luxurious age and time; but there will be and there are churches for the average laborer; churches glad to see anybody who is decently clad, and to give a good seat to the man who may be hungry and possibly not quite cleanly. I believe that nine out of ten of our churches are willing to see all ranks of society in God's house, and to measure them there only by the standard of religious character. When the classes that we wish to reach are not reached by the regular churches, and when they can be reached by tabernacles and Young Men's Christian Associations; when an audience of five thousand comes together in an open hall, — can such an opportunity be innocently thrown away?

We are, I think, far underrating the willingness of the rougher class in our large cities to hear Christian truth. We are far from meeting their hunger. The intensity of desire on the part of hundreds and hundreds who have given up hope, to be encouraged, to be told that there is yet a prospect for them, although they have not where to lay their heads, is greater than you imagine. You do not go down into the lower strata of society. You sit before your fender; you toast your moccasons there; but if you would stain them a little in the gutter, and in the rough straw of the attics, and in the damp mire of the cellars, where more and more of our population in cities are living, you would find yourselves on the path followed by Him who went about from house to house doing good. [Applause.]

THE LECTURE.

At the Diet of Worms, Martin Luther, when requested to recant, began the modern discussion of conscience by saying, "Here I stand. I can do no other. It is not safe for a man to violate his conscience. God help me!" In these words, Protestantism put her foot upon a piece of granite, which modern scientific research is now convinced takes hold on the core of the world. Theology, in that speech of Luther's, took its position upon self-evident truth in regard to the moral sense, and asserted three things: —

1. That a man has conscience.
2. That God is in it.
3. That it is not safe to disobey a faculty through which God looks, as of old he looked through the Egyptian pillar of cloud and fire in the morning watch, troubling the hosts of all dissent.

More and more fruitfully, since Luther's day, religious investigation has taken up the topic of conscience from the point of view of the scientific method. Bear with me, my friends, if, in discussing conscience as the basis of the religion of science, I take you over definitions which may appear at first dry, but out of which, possibly, may germinate umbrageous foliage in which the very birds of heaven may sing, and under which at last we, in the dust and heat of these tempestuous days of debate, may sit down in peace, and be refreshed.

1. Sensation and perception always co-exist.
2. Sensation involves perception first of the sensation or feeling itself, and second of an object causing the feeling.
3. The intensity of sensation and that of perception, when both are exercised at the same instant, are in an inverse ratio to each other.

4. These are the laws of touch, taste, sight, and all the physical senses.

"Knowledge and feeling, perception and sensation," says Sir William Hamilton (*Lectures on Metaphysics*, p. 336), " though always co-existent, are always in the inverse ratio of each other. That these two elements are always found in co-existence, is an old and notorious truth."

It is sometimes asked how I can possibly define conscience as both a perception and a sensation. We perceive the difference between right and wrong intentions. We feel that the right ought to be chosen, and the wrong rejected, by the will. Both these acts, I affirm, proceed from conscience. A being incapable of either act we could not say has a conscience; and this proves that both the powers must be named in any definition of conscience. But here are two opposite activities, some say. Must not conscience be either all intellectual or all emotional? Is it not all a perception or all a feeling? What is conscience in the last analysis, perceptive or emotive? Suppose that you ask this question concerning the sense of the ludicrous, or that of the beautiful. Each of these plainly includes both perception and feeling, as does conscience.

5. The sense of the beautiful involves a *perception* of the distinction between beauty and deformity, and a *feeling* of delight in the one and of distaste for the other.

6. The sense of the right involves a *perception* of the distinction between good and bad motives, and a

feeling of delight in the one and of distaste for the other.

We must not confuse together conscience and taste, the moral and the æsthetic, the sense of the right and that of the beautiful; but there are most subtle and significant resemblances between the laws of these two faculties. I have some strange object presented to me, and I perceive it, and I feel at once that it is either ugly or beautiful. A crooked line, a gnarled, jagged figure, is not as beautiful as a circle. If you attack me here, I can only reply that these are self-evident truths concerning beauty and taste. I have a sensation; and connected with that sensation is a perception of beauty or deformity. The sensation of your gnarled, jagged line gives me a perception of what I call deformity, and the sensation of the circle gives me a perception of what I call beauty. So too the sensation within my soul of a motive which is not harmonious with all the light I possess gives me the impression of moral ugliness; and the sensation of a motive perfectly conterminous and harmonious in all particulars with the best illumination I possess or can obtain, gives me an impression of moral beauty. Jonathan Edwards described virtue as the love of right motives considered as morally beautiful, or as admiration for goodness as beauty of a spiritual sort.

7. The perception and feeling and love of æsthetic beauty are pleasurable.

8. The perfection and feeling and love of moral beauty are blissful.

Thus the question as to whether the sense of right is feeling or perception is answered by attention to analogy and fact. Sensation implies perception. The sense of the beautiful includes both perception and feeling. It is not proper to ask concerning the sense of the beautiful or that of the ludicrous, whether it is intellectual or emotional. Each is both; and the sensation involves the perception. Just so the sense of right involves perception necessarily. So, also, in my power of physical touch and taste, sensations involve perception.

9. By physical sensation and the involved perception, we have a knowledge of physical realities outside of us.

10. By æsthetic sensation and the involved perception, we have a knowledge of æsthetic realities outside of us.

11. By moral sensation and the involved perception, we have a knowledge of moral realities outside of us.

12. All the certainties of physical science depend on the trustworthiness of the self-evident truths visible to us in the perception which is involved in physical feeling.

13. All the certainties of æsthetic science depend on the trustworthiness of the self-evident truths visible to us in the perception involved in æsthetic feeling.

14. All the certainties of moral science depend on the trustworthiness of the self-evident truths visible to us in the perception involved in moral feeling.

FOUNDATION OF THE RELIGION OF SCIENCE. 215

15. The three classes of certainties, — physical, æsthetic, and moral, — as depending equally on self-evident truths visible to us in perceptions involved in natural sensations, are of equal degrees of authority.

16. The ultimate tests of certainty in physical, æsthetic, and moral science, are therefore the same in kind.

When I take in my hands any physical object, I in the first place feel it, and am conscious of the sensation; in the second place, I am sure that something is the cause of that sensation, and that the something is not myself. It is outside of me. There is the beginning of the range of sensation. This feeling involves perception, not of all the qualities in the external object, but of the fact that there is an external object. I do not know what is in a book by touching it, but I know that I touch somewhat, and that the somewhat is not myself. It is so in sight and in hearing. I am conscious first of the affection of my own personality, and then of a something outside of myself causing that impression. I have no control over the laws governing physical sensation.

Just so, rising into the range of taste, I find that the laws of beauty are not ordained by myself. I see what I call ugliness, and I *cannot help* finding it distasteful. I see what I call beauty, and I *cannot help* having a delight in it. That law of distaste or of delight is not subject to my will. It is above me. I feel that it is something outside of me, and that it has authority in the universe without my consent. It is one of the laws of things, just as much as the law of gravitation.

We are all agreed up to this point. We have an experience of sensation involving perception of the law of physical gravitation. We do not know all about it, but what little we do know concerning it is sure as far as it goes. Just so I do not know all the laws of the beautiful, but I know that there is a distinction between deformity and beauty, and that this distinction is outside of me, and in the nature of things. As by the evidence of the physical and æsthetic senses I find out that there is a physical law of gravitation outside of me, and that there is a law of beauty outside of me, so, when I rise into the higher faculties of the soul, I find that they have sensations, and that their sensations involve perception, and that yonder, in the loftiest part of the azure of the sky within us, there are laws, just as surely as in this mid-sky or the region of taste, and just as surely as upon the earth on which we tread. Here are physical things — sensation involves perception; here are æsthetical things — sensation involves perception; just so there are moral things, and sensation there, as elsewhere, involves perception. Therefore if you follow the scientific method based on the trustworthiness of your sensations and the involved perceptions in physical things, and follow the same method based on the trustworthiness of your sensations and the involved perceptions in æsthetical things, I will go farther, and affirm in the name of the universality of

FOUNDATION OF THE RELIGION OF SCIENCE. 217

law precisely what you have affirmed over and over again, namely, that sensation involves perception; and I will apply this principle to moral as you have to physical and æsthetic perception; and thus I will find in the upper sky a law by the scientific method, just as we find one in the mid-sky and on the earth. [Applause.] If objective reality is guaranteed by a constant experience in the one case, it is in the other.

17. We have a constant experience that our natures are made on such a plan that we distinguish between rightness and wrongness in motives.

18. We have a constant experience that we are made on such a plan that we feel irresistibly that we ought to follow right motives, and not follow wrong.

19. We have a constant experience that pain or bliss follow duty neglected or duty done.

20. We have a constant experience that a sense of an approval or disapproval higher than our own follows duty performed or duty disregarded.

21. We have a constant experience that our faculties forebode our personal reward or punishment in another state of existence, according as we do or do not follow conscience.

22. The constant experience of moral sensation and perception is as perfect a ground of certainty as to moral law as a constant experience in æsthetic sensation and perception is in regard to æsthetic law, or as a constant experience in physical sensation and perception is in regard to physical law. (See a fresh,

keen book by Newman Smyth, *The Religious Feeling.* New York: 1877.)

It is a suggestive remark of Nitsch, the great German theologian, that "the religious consciousness perfects and justifies itself, when, in the immediate life of the spirit, what is contained in the original feeling of God (Gottesgefühl) objectifies itself in a *constant* manner." (*System der Christ. Lehre*, p. 25.) The far-reaching law that a constant experience is the guaranty of all scientific certainty bears all the tests applied to truth within the range of physical investigation. Your Tyndall, your Huxley, your Spencer, have in physical science no grounds of certainty that do not depend upon a uniform physical experience. We have dreams, to be sure, in which certain strange things occur to us; but the dreams proceed according to laws which are not a constant experience. We find that they lack verification in other positions of our consciousness. We are not always treated by the external world as we are in dreams. But when we, as individual men, and waking, have a constant moral experience; when, age after age, we as a race walk waking through all the environments of history; when age after age we walk waking under all the winds that beat upon us from out of the skies of moral truth; when we find constantly that there is a difference between right and wrong, and that we feel we ought to follow good motives, and not follow bad; when constantly we are beaten upon in the same way,— then these impressions made upon us are revelatory of the moral plan, not only of our natures but of our environment, and the constancy of moral experience is to be looked on as is the constancy of æsthetical and the constancy of physical experience, as a source of scientific knowledge.

Pardon me, my friends, if I say that modern scepticism appeals to Cæsar, and to Cæsar it shall go. [Applause.] You believe, you say, and you adhere unflinchingly to all self-evident propositions within the range of physical research. Sir William Hamilton and Kant and many another philosopher have divided our faculties into the understanding and the reason. By the reason, as understood by Kant, we do not mean the understanding, but *the faculty of perceiving self-evident truth*. Now, there are self-evident truths in the range of morals as surely as in the range of physics. Kant's practical reason, or faculty by which we perceive self-evident truths of the moral kind, is only another name for conscience, or the moral sense. There are self-evident truths in the range of æsthetics as surely as in the range of morals. We have a faculty by which we perceive self-evident truth; or, rather, our whole nature is so made that we cannot but believe self-evident propositions. Look for a moment at these different lists of propositions. Take a few merely intellectual self-evident truths, such as the geometrical and mathematical axioms. We are all convinced, not merely by evidence, but by self-evidence, that the whole is greater than a part, and that two straight lines cannot enclose a space, and that every change must have a cause. Just so in the range of æsthetics, although the intuitions there never have been as carefully studied as in the range of mathematics, we are sure that there is a difference between beauty and deformity. We do perceive by direct vision that a circle and an ugly gnarled line are different, and that the one must be put on the right hand and the other on the left before any judgment-bar of taste. All men agree in these feelings, and say the self-evident truth involved in them is that there is a distinction between the right hand and the left in every thing touched by our sense of the beautiful. But we rise into the region of morals, and there is yet greater clearness than in the region of taste. Here is an intellectual axiom, you may say, but it is really a moral one: Sin can be the quality of only voluntary action. There is a perfectly self-evident moral truth. You cannot prove it by any thing that does not assume it. It is not only evident, but it is self-evident. It is a moral axiom, and you are just as sure of it as that two and two make four. Sin is free, or you cannot make sin out of it.

Tyndall now publicly agrees with Häckel in maintaining that the will is never free. Echoes are already beginning to be heard, even in Boston, of his Birmingham assertion that the robber, the ravisher, the murderer, offend because they cannot help offending. They are to be punished, indeed; but they are no more blameworthy than honest men and reformers and saints and martyrs are praiseworthy. In this city I read in an editorial yet wet from the press the assertion that the criminal offends because he cannot help offending, and that such a doctrine permeating society would free us from a large amount of theological quackery. Will the teachers of this atrocious shallowness insure the prisons against the effects of their own quackery? Will they lift off from trade and social life the weight of this false science, which, if trusted, will ride greed and fraud as never nightmare rode invalid? When the last word of the Häckelian evolutionists, — opposing Darwin, opposing Dana, opposing Owen, opposing every anti-materialistic theory of evolution in England or Germany, and all similar schools in metaphysics, — is a denial that the will is ever free, and an assertion that the murderer and the robber and the ravisher offend because they cannot help offending, it may be said with justice that the materialistic cuttle-fishes are trying to attack the leviathians of self-evident truth, by throwing off ink into the sea! They will succeed in making things clear only when the sea is all of their own color.

If a man is to be loyal to axioms, if a thinker is to require of himself consistency, if there is to be clearness or straightforwardness in thought, we must demand that the scientific method, rising thus from the physical to the æsthetical, and into the moral, shall hold fast to self-evident truth yonder, just as in the mid-sky and on the sods of purely physical research. I will not admit that the whole world belongs to the men who follow scientific truth only in its physical relations. Heaven forbid that I should deny that they are making important discoveries!

They mine far into the earth, they sink wells down and down; but at the bottom of their wells, looking upward, they do not see the whole range of truth. It is important to recognize the merit of men who sink wells into the earth; but if they, as specialists, are to have sound minds, they must come often to the curb-stone, and at least put their heads out, and gaze around, north, south, east, and west. (SMYTH, *The Religious Feeling*.) They will find the mid-sky a fact, as well as the bowels of the planet; they will find the upper sky a fact, as well as the mid-sky, and as well as that inner vein which they have been working. We are not out of the range of gravitation when we are out of the physical specialists' well. We are not out of the range of self-evident truth when we rise out of the mine, and look around us and above us. Forever and forever, we must acknowledge the unity and the universality of law; and therefore self-evident moral truth will be to us always a pedestal from which the philosophy of religion will be visible to its very turret, if only we carry up her telescope to that summit along the line of the only rent through the clouds that God's own hand seems to have made when he stretched forth his creating arm, and implanted these self-evident truths in the human constitution. [Applause.]

23. We know incontrovertibly, therefore, by a constant experience of a moral law and of a Personal Power not ourselves, that makes for righteousness, that the plan of our natures, taken as a whole, and the environment we have here and hereafter, require us to choose what ought to be.

24. But to choose is to love.

25. To follow the plan of our natures, or conscience, both in what it includes and in what it implies, we must therefore love a personal God, revealed through the imperative commands of conscience, and in the pains and blisses of our constant moral experience.

If any one scheme of philosophy now appears more likely than another not to disappear, it is that of which the fundamental thought is an ethical representation of the universe. The philosophy of Lotze, like that of Leibnitz and Plato, turns on the central principle that the ends of the universe are moral. One of Lotze's profoundest sayings is: "The world of worths is the key to the world of forms." This is the deepest ethical teaching of your Julius Müller, and of your Dorner, your Rothe, and your Ulrici, — that we never understand any thing until we connect it with the moral purpose had in view by the Author of all things from the first. Study physical science only, and perhaps you may be tempted to conclude, as Stuart Mill did, either that God is limited in power, or that there is a doubt of his goodness. But when we turn from external nature to the moral law, revealed by the scientific method; when we fasten our attention upon the great tendencies and influences which are to give ethical causes supremacy, and make the right victorious; when we remember, with Matthew Arnold, that the Eternal Power which is outside of us makes for righteousness, and makes imperatively for it, and victoriously for it, — we see that the end is not yet; that the scheme of the universe is not fully executed; that the perfection of the moral law prophesies the perfection of the ultimate arrangements of things; and that, therefore, in conscience we have an observatory higher than that of physical science ever was, from which to gaze upon the supreme harmonies of the universe.

He who enters into the depths of his conscience, and there muses, pacing to and fro, is more likely to meet God, and to understand the plan of the whole universe, physical as well as moral, than he who paces to and fro among the Seven Stars, or puts his hand upon the sword-hilt of Orion, or flies with Cygnus across the meridian, or follows Boötes as he drives his hunting dogs over the zenith in a leash of sidereal fire. He who fastens his attention on the uppermost ranges of natural law will understand the lower, into which the upper sink down with supreme power. He who gazes only upon the planets will understand neither the planets nor the suns. Begin with the loftiest that is known to us; take the scientific method up into the constellations which in all ages have had constant forms in the human inner sky; study the sense of dependence and obligation which point to a personal God, — and you will find that the universe has everywhere an ethical tendency; you will find that the ethical aim of all things is the justification of all things, and in conscience will discover the Copernican system of the moral heavens.

"Love God," writes Thomas Carlyle: "this is the everlasting Yea in which all contradiction is solved;

and in which whoso walks and works, it is well with him." (*Sartor Resartus.*)

Repetition of experiment! That is the scientific test of deepest significance. Religious Science does not flinch in the application of it. In that test she finds all her victories. She asserts that there is a Power that makes for righteousness, and points to all history as a repetition of tests of that truth. She asserts that conscience crowns whoever yields to its demand of personal self-surrender to the moral law and to the personal Lawgiver revealed through moral sensation and perception. Her assertion she justifies by repetitions of experiments in individual lives, age after age. The more perfectly you adhere to experiment, the more are you fortified in belief of all the great truths concerning conscience. Who are these sceptics who revere the scientific method, and are unwilling to try experiment even once concerning this upper realm of truth? I assert that it is a fixed natural law, that when the soul yields utterly to God He streams into the spirit, gives a new sense of His presence, and imparts a strength unknown before. Will you try such self-surrender, and then will you repeat the experiment as opportunity offers, — I care not how often? Every path of choice divides before me. The right hand or the left I must take, and I take the right. Immediately the path divides right and left again. I take the right. Immediately it divides. Every choice as to the path has a moral character; and so either sin rolls up fast, or the habit of virtue grows fast. Every day you put forth billions of choices, and in every choice there is a moral motive. But now I affirm that in these billions of opportunities for experiments, in these ten thousand times ten thousand chances to test whether I am right or wrong, you will not find one chance failing to give you this verdict, that, if you yield utterly to God, he will stream through you. Whenever your conscience is made gladly supreme, its yoke, by irresistible natural law, will transform itself into a crown. This constant experience you will have at every forking of the ways; and so every forking will be to you, if you choose to make it such, a repetition of experiment, and a verification of the trustworthiness of the scientific method applied to the innermost holiest of the soul. Rising through that constant experience, we may, even in our present low estate, approach the bliss of the upper ranges of being, and of those who never have sinned, and of that Nature which was revealed on earth once, as the fulness of Him who filleth all in all. His bliss is the brightness of all infinities, and is symbolized to us by our own intellectual, æsthetic, and moral gladness, when we are right with a universe in which all law is one thought, and that His own. It should be asserted by science in the name of experiment, that man may become a partaker of the Divine nature. Adjust the conscience to the law it reveals, and He whose will the law expresses will invariably produce in the soul the largest measure it can receive of his own bliss and strength. [Applause.]

IX.

THE LAUGHTER OF THE SOUL AT ITSELF.

THE EIGHTY-NINTH LECTURE IN THE BOSTON MONDAY LECTURESHIP, DELIVERED IN TREMONT TEMPLE, NOV. 26.

O die Wunde des Gewissens wird keine Narbe, und die Zeit kühlt sie nicht mit ihrem Flügel sondern hält sie blos offen mit ihrer Senfe. — RICHTER: *Titan*, cycle lxxxii.

> Νόμος ὁ φυλαχθεὶς οὐδέν ἐστίν, ἢ νόμος,
> ὁ μὴ φυλαχθεὶς καὶ νόμος καὶ δῆμος.
> MENANDER: *Ex Legislatore*.

IX.

THE LAUGHTER OF THE SOUL AT ITSELF.

PRELUDE ON CURRENT EVENTS.

POOR Jean Jacques Rousseau affirms in his Confessions, that the happiest instants of his entire career, which ran, as you know, through unreportable villanies and leprosies, were in an afternoon, when he was yet virtuous, and met a company of young people, themselves yet virtuous, and felt the strange power of the pure atmosphere that comes into the world with us, as he breathed it deeply in the height of sociality. A thoroughly atheistic Frenchman, who lived on the whole a life less cleanly than that of a beast, said in his mature age that if he could have known in his youth what kind of a time he was to have, he would have hung himself.

There is in Boston a quarter which the sailors call the Black Sea. Not every one there is as wise as Rousseau, or as this atheistic Frenchman; but of course Boston is as wise. On many a shore of the ocean, seaports with Black Seas in them exist. If Boston could once show her wisdom by making cleanly this one Black Sea, she would set an example for all coasts. We draw near Thanksgiving morning, and have I not a right to speak of Magdalen in cities? I know on what ground I am treading, and that if any speaker slips here, there hangs over him the crack of doom; but one purpose of this lectureship is to discuss themes that cannot well be noticed elsewhere. If you will bear with me, I must say that Boston has as noble facilities as any city on the globe for solving the problem of the management of the corrupt and perishing classes in great towns. More than half the population of this Commonwealth live in cities. Boston is not so painfully under the control of a foreign-born vote as is the metropolis at the mouth of the Hudson. In New York two hundred thousand of the inhabitants were born in Ireland, and one hundred and fifty thousand in Germany. I know that New England, in its manufacturing centres, is becoming New Ireland; I know that Boston, within municipal limits, is becoming an Irish city. But take Boston sleeping-rooms into view, or the circuit of the fifty miles in each direction in sight of the State House, and the population within that space is as American and as enlightened as that in any other quarter of equal size on the globe. Property is more equally diffused here than in any other section of equal extent; and so are intelligence and virtue. I know what large claims these are, but I am not a citizen of Massachusetts by birth. I am proud of my native State, New York, with the great Sound and the gates of the ocean at one corner of its wide territory, the Adirondacks at another, and the eternal roar of Niagara at a third. But you have opportunity here, which New York may never possess, to wash a desolate city quarter white, or at least gray. If you do not improve the opportunity, the time will come when even Beacon Hill will be aware of the presence of the Black Sea in this municipality. The two quarters are not far apart, a small fraction of a mile, and yet they are not acquainted with each other. In many seaports of the world the Black Sea and the Beacon Hill exist, but they rarely understand each other. Is it not time, now that God is massing men in great towns, and especially in seaports, for Americans who claim to have political ingenuity and moral enthusiasm, to ask whether there can be a noose made that will throttle the enemies of Magdalen?

What can be done for Magdalen in cities? Seven things.

1. Visitation of the degraded is possible to women.

This remedy of personal intercourse with those who have gone down beyond the lowermost round of the ladder that leads into society is a two-edged method of action. In the first place, it teaches the haughty and the luxurious who go down there that life is not all of the smooth sort, and that really, in this nineteenth century, and to the last hour of the unrolling ages, there are places into which men cannot venture safely, and especially not women. In all velvet society we need to be taught that between the right hand and the left there is a difference absolutely infinite. The chief merit of the measure of personal visitation is in its reflex action upon a luxurious, soft, hammock-swung, lavender Christianity.

You have here in the North End, close under your windows, children that are born cherubic, possibly, but who grow impish very fast. They are elbowed by the dance-hall. They look out of their cradles into brothels. Behind their nursery windows stand the reeking stables. Up and down the gutters stagger and fight men whom drink has made demons. They curse each other in the hearing of the young ears. Women whom drink has made furies preside at many cradles. Sottish and leprous parents ought to perish, you think; but what of their children? The shiftlessness of the Portuguese and the Italians and the Irish, and the nineteen other nationalities who are represented in that Black Sea, deserves the spur of hunger, you say. But are the children to blame for being there? Have they not a right to a permanent place in your pity? Surely they did not choose the spot in which they should come into the world. After all that we say haughtily about letting vice take its own course, we must remember that children start weighted in the race of life, and that we ourselves put upon them some of the weights if we allow these desolate quarters to go without religious, social, and financial visitation.

2. The opening of homes for the degraded is possible.

3. The sending of the reformed out on demand into families is possible.

You believe in experience. I hold in my hands official statements which are authorized by some of the noblest signatures in the city, and which I might justify by giving names. On the authority of these statements I assure you that it is a fact that some graduates of homes for the fallen are now members in good standing in Christian churches in this city and vicinity. It is a fact that several of them have so comported themselves in the households where they have been placed, that intelligent clergymen and clear-sighted matrons have written in the highest commendation of them. It is a fact that some have scarcely wavered for ten years, and that then the open bars of our city were the pitfalls which caused the temptation. It is a fact that at least one of the homes which these official authorities represent has had more demands for graduates from it than it could supply. I am speaking of the Mount Hope Home, if you will have me be definite, in charge of the North End Mission — no sectarian enterprise. I do not underrate the numerous priceless denominational enterprises in this Black Sea. I speak for them all in speaking for the North End Mission, which is aided by all denominations. This home, supported by that mission, is a staircase up which degraded persons have ascended, — helped by the angels, no doubt, — and have reached the highest standing-place in some cases; have had opportunity to offer themselves to God; have escaped from the Gehennas of this life. And now there is a bar across that staircase! What is it made by? I look at it with amazement. I can hardly believe it is a bar of any thing but the vapor of the harbor. I can hardly believe it is any barrier to the ascent of these degraded ones to a life of reformation. I come nearer to that bar. I look. There is an inscription on it. What is it? "Shut for want of funds." And underneath is written "Boston"—is it "Boston penuriousness," or "Boston carelessness"? There is a fog there; I will not try to read the inscription: it is one or the other. [Applause.]

You are setting an example, are you, for all the seaports of the world? When official testimony of this kind is put before you; when little Boston, easily managed, if men make up their minds to do their duty, is thus in a strategic position among American cities; when New York and San Francisco and New Orleans, and Liverpool, and Lisbon, and Naples, and all the Black Seas the world around, are each throwing up to the sky a glance like a gleam of light out of a serpent's eye, and you are asked here to put out one of those eyes once for all, and change one Black Sea into a sweet pool of waters, you fold your hands; you say that these things must take care of themselves, and that the whole problem perhaps is insoluble. And yet those who go down into these dark waters, men who have made specialists of themselves there, and some of them are highly educated, assure you that nothing is needed but financial and moral support to secure again and again a passage up that now blocked staircase for those whose feet and bodies and whole form to above the lips — they cannot call out, they have no voice, and I give them what little voice I have — are submerged. You say these men are wild; but they say that those who are sunk even beyond the lips and even beyond the eyes, and cannot see their own condition, may emerge, and put on white robes. [Applause.]

4. There may be execution of law against houses of death.

You vote for mayors and aldermen; you have serious views as to how this city ought to be managed. You are intending to reform it by a paper constitution by and by. You are determined to have a responsible mayor in this city. The lack of an executive that can be brought to justice is, indeed, the chief deficiency in our municipal governments throughout the United States. But the people are mayors; the people are aldermen. The careless voting of American cities, when attempts are made to avoid the execution of the law, is something that ought to make the statues of the fathers here in Boston leap from their pedestals.

5. There may be laws to hold men to as stern an accountability as women on the public streets. [Applause.]

6. The temperance laws may be executed.

7. The German social standards in pagan days may be revived.

What does Tacitus say of our fathers, when, under the German forests, they were first brought within range of the historic telescope? They were monogamists. The love of home was one source of the patriotism of the Teutonic tribes. The Romans never conquered our fathers. Is the love of home likely to be undermined among Anglo-Saxons? Did you read Herbert Spencer's Sociology? Did you not find him turning all the light of advanced thought upon the question which lies at the centre of social life; and justifying, in the name of philosophy of the freest sort, the soundest ideas on that theme? Perhaps, if you will be as anxious as Spencer is, to understand natural law, you will agree with him thoroughly in his organizing and redemptive conclusions concerning sociology. You know that I am not a eulogist of Spencer in general; but he has said lately a few things which look wiser than his earlier declarations. (SPENCER, *Principles of Sociology*, 1876, vol. i., part iii.) The truth is, that the family is more and more put in peril by the advance of luxury in civilization and by the massing of men in cities, and by a leprous philosophy that holds that man is never to blame, whatever he does. Are we to sit still, and have that doctrine taught? Are we to let the trail of that serpent drag itself over the leaves of the vines that cluster on the trellis-work of our homes? Herbert Spencer sends out no such creeping worm of the Nile into social life. Materialistic philosophers have done so lately.

There are many Saxon faces in this audience. The blue eyes, the white forehead, the blonde cheek, the fair hair, are signs of the Anglo-Saxon lineage. That race rules the world to-day. It may not always rule it. It rules it for a cause. That race has given to us Goethe and Milton and Shakspeare; and Bacon and Kant and Hamilton and Edwards; and Cromwell and Washington and Lincoln. It wrote Magna Charta, the English Constitution, the Declaration of Independence, the Constitution of the United States. It has bridged the ocean with its commerce, and traversed it with its electric wires. That race, in its German forests, was noted for nothing so much as the spotlessness of its private morals. While yet barbarian, our German fathers, as the Roman historians state, buried the adulterer alive in the mud. The adulteress they whipped through the streets. "*Non forma,*" says Tacitus, "*non œtate, non opibus, maritum invenerit.*" "Neither beauty, nor youth, nor wealth, found her a husband. They considered," Tacitus says, "that there was something divine in woman, and that presaged the future; and they did not scorn her counsel and responses." Youth were taught chivalric notions of honor. Out of this race sprang chivalry. It is this race which has proved itself, in the hurtling contests of a thousand years, both in peace and war, superior to all relaxed Italian and French tribes as the leader of all the world's civilization. The purity of the tribes in the German forests prophesied their future. The hiding of the power of the Anglo-Saxon race has been in the fact that it was at the first free from the sin of Sodom and Gomorrah. That race is passing the trial of power. It is passing the trial of luxury. In the German wilds our fathers, as the Romans found them, were, as a race, as pure as the dews the forests shook upon their heads. The race has predominated in history, because free, even when barbarian, from what elsewhere has been the commonest leprosy of barbarism. It will continue to predominate if it continues free. If the Anglo-Saxon race has shown exceptional vigor, the chief secret of its power is to be found in its reverence for a pure family life. [Applause.] It will continue to have power, and rule the world, if it continues that pure life; otherwise, not. [Applause.]

THE LECTURE.

The innermost laughter of the soul at itself, it rarely hears more than once without hearing it forever. What is the laughter of the soul at itself? Do you not know, and do you wish me to describe, this convulsion of irony, of fear, it may be of despair, which sends cold shivers through all our nerves, causes a strange perspiration to stand on our foreheads, and makes us quail, even when alone — as we never are? You would call me a partisan, if I were to describe an internal burst of laughter of conscience at the soul. Therefore let Shakspeare, let Richter, let Victor Hugo, let cool secular history, put before us the facts of human nature.

Here is Jean Valjean, principal character in Hugo's *Les Misérables*, one of the six best works of fiction the last century has produced. Hugo is no theologian. He is not even a partisan teacher of ethics. He is a Frenchman. His ideals have been obtained largely from Paris. But you open his chapter entitled "A Tempest in a Brain," and you find him asserting that "there is a spectacle grander than the ocean, and that is the conscience. There is a spectacle grander than the sky, and it is the interior of the soul. To write the poem of the human conscience, were the subject only one man, and he the lowest of men, would be reducing all epic poems into one supreme and final epos. . . . *It is no more possible to prevent thought from reverting to an ideal than the sea from returning to the shore.* With the sailor this is called the tide. With the culprit it is called remorse. *God heaves the soul like the ocean.*" Elsewhere this modern Frenchman writes: "Let us take nothing away from the human mind. Suppression is evil. Certain faculties of man are directed towards the Unknown. The Unknown is an ocean. What is conscience? The compass of the Unknown." (*Les Misérables*, chapter entitled "Parenthesis.")

Valjean here has been in the galleys. He has escaped, assumed another name, and has become the mayor of a thriving French town. In his business he acquires the respect of all who know him. But one day, an old man who has stolen a bough of apples, and who looks like Jean Valjean, is arrested as Valjean himself, and is in danger of being condemned to the galleys for life. There is a striking resemblance between the faces of the two men. The true Valjean is brought face to face with the question whether he will confess his identity, or allow another man to go to the galleys in his place. Valjean has tried to recover his character. A bishop, who taught him religious truth, seems to hover in the air over him. A couple of golden candlesticks which the bishop gave him, he treasures as possessions priceless for their reminiscences. He goes to his room; shuts himself in; and, as Victor Hugo affirms, he was not alone, although no other man was there. Valjean meditates on his duty, and his mind becomes weary under the tempest of conflicting motives. Shall he go back to the galleys? Shall he be whipped up the side of the hulks every night in loathsome company? Shall he feel the iron on his ankles and on his wrists? Shall he hear nothing but obscenity and profanity the livelong, hard-working day? Shall he give up the opportunity of being a benefactor to a wide circle of the poor? Ought he not to make money, that he may give it away? We have forgers who ask that question. [Laughter.] It is said that some men have thought it a convenient modern trick in trade, to endeavor to persuade one's self that the infinite weight of the word *ought* lies on the side of philanthropic forgery. But Victor Hugo does not represent Jean Valjean as of that opinion. In spite of all the temptations found on that side, Valjean at last concludes that it is his duty to declare his identity, and save this Champmathieu from the galleys.

But then, as you remember, there comes another thought to Valjean. Fantine, a ward of his, and her child Cosette, depend on him exclusively. The mother has suffered nearly every thing, and deserved to suffer much, but without Valjean her life and that of her child will be a ruin. "Is it not," he asks, "a clear case that this old man, who has but a few years to live, is worth less than these two young lives?" Throwing himself out of the case, Valjean must leave either him or them to fate. Reasoning thus, he at last adds his former selfish temptations to these unselfish ones. He remembers his duties to himself and his duties as a benefactor. He sums them all up; and says that, after all, nobody knows that he is Jean Valjean. He has only to let Providence take its course. God has decided for him. He makes up his mind not to declare himself. "Just there," Victor Hugo says, "he heard an internal burst of laughter." Hugo affirms that a man never hears the deepest laughter of this kind more than once, without hearing it during his whole existence, here and hereafter.

Valjean, however, persists in his resolution not to declare himself. He repeats his reasoning in self-justification; he thinks that he speaks from the depths of his conscience; "but still he *felt no joy.*" This sign of self-deception does not induce him to pause. He takes down his old galley suit, burns it; finds the thorn stick, with its iron-pointed ends, which he had used when a vagabond, burns that; gazes on a coin which he robbed from a boy, puts that in the fire; and finally he prepares to destroy the two golden candlesticks, which years before were given him by the bishop, who now seems to be in the air at his side, not able to face him quite, but whispering behind his ear. He takes these candlesticks, bends over the fire, almost stupefied by the violence of his emotions; warms himself at the crackling flames; throws them in — "Valjean!" He looks up, and there is no one present. There was some one there, Hugo says, but He was not of those whom the human eye can see. "Do this," continued the voice, which had been at first faint, and spoke from the obscurest nook of his conscience, and which had gradually become sonorous and formidable, and seemed to be outside of him: "put into the flames all that suggests reminiscences of the devout sort. Make yourself a mask if you please; but, although man sees your mask, God will see your face; although your neighbors see your life, God will see your conscience." And again came the internal burst of laughter: "That is excellently arranged, you scoundrel!"

Midnight struck. Valjean heard two clocks. He compared the notes, and he was reminded that he had seen a few days before, in a shop, a bell having on it the name Romainville. Hugo is a subtle poet. He says much between the lines. Suddenly Valjean remembered, says Hugo, "that Romainville is a little wood near Paris, where lovers go to pick lilacs in April." Valjean falls asleep, and has a dream. He is near Romainville, but all the houses are of ashen color; all the landscape is treeless and ashen; the very sky is of leaden hue. He enters Romainville, where the lilacs grow that the lovers pick in April, — deep allegory this, by a Frenchman, no partisan, no theologian, — and around a corner where two streets meet, he sees a man leaning against the wall. "Why is this city so silent?" The man makes no reply. Valjean enters a house. The first room is empty; in the second room, behind the door, he finds in his dream another silent man, leaning against the wall. He asks him why the house is deserted, but no reply is given; and all the walls are ashen color, and the sky continues to be leaden. He wanders into house after house. He finds a fountain bursting up in a garden, and behind a tree a man, but he too is silent. There was behind every corner, every door, and every tree, a man standing silently. Before entering Romainville, he met on the plain near the city a horseman, "perfectly naked," — Hugo writes, and he knows what he means,— and with a skull instead of a head, but yet the veins were throbbing around the skull; and in his hand there was a wand, supple as any grape-vine, yet firm and heavy as lead. With that wand this horseman was to chastise the inhabitants of this city. Valjean, in his dream, went out of the lifeless town in horror, and, looking back, he saw all its inhabitants coming after him. They saluted him on the open plain, under the leaden sky, and this was their language: "*Do you not know that you have been dead for a long while?*" Men who have heard the internal burst of laughter as forgers, as lepers, as those who dare not open their souls to their neighbors, find behind the doors and in the booths, and even on the street-corners, silent men; and when these criminals, known to God under their masks, walk into solitude, those silent men come after them; and, when once conscience has been finally insulted, the cry of all the nature of things is represented by the inhabitants of Romainville in Victor Hugo's dream. Instead of lilacs in April, you have the leaden sky; you have all the earth dun-color; you have a brazen sod on which to stand; you have this horseman, with the whip lithe as a grape-vine and heavy as lead, before you; and behind you this host with the cry, "Do you not know that you have been dead a long while?" [Applause.]

Valjean finally confessed his identity; and the court and audience, when he uttered the words, "I am Jean Valjean," "felt dazzled in their hearts," Hugo says, "and that a great light was shining before them." [Applause.]

Take Richter's Titan, another of the six greatest works of fiction the last century has given to the world, and perhaps the greatest of them all. Roquairol, the fiend of the book, dies by suicide. He utters no words which the Titanic Richter, no partisan, no theologian, does not put into his mouth. Richter's human horologes have crystal dial-plates and transparent walls which allow us to see the mechanism within. More than once this Roquairol has heard the laughter of his soul at itself. "I cannot repent," says the leper, with his pistol at his own brain. "Should that which time has washed away from this shore cleave again to the shore of eternity, then it must fare badly with me there. I can change there as little as here. I do verily punish myself, and God immediately judges me." Here he suddenly points the weapon at his forehead, fires, and falls headlong; blood flows from the cloven skull; he breathes once, and then no more. Albano, the serene, vast soul which represents Richter's views of conscience, stands at the side of the corpse, and seems to hear the words from the suicide's breast and iron mouth, "Be still: I am judged." (*Titan, Cycle,* 130.)

But, you say William Shakspeare would not be as melodramatic as this Frenchman Hugo, nor as serious as this German Richter. He was an Englishman. Although Tennyson has lately praised Hugo in a sonnet, and although Mrs. Browning has said that Dickens learned to write fiction from Hugo (*Letters of Mrs. Browning,* vol. ii.), you will follow no French authorities as to conscience. John Calvin was a Frenchman [applause], and did not teach fatalism either. [Applause.] Shakspeare more than once has represented the despair of the soul under the law of its own nature: —

> "Oh, my offence is rank, it smells to heaven!
> It hath the primal eldest curse upon it,
> A brother's murder. Pray can I not,
> Though inclination be as sharp as will:
> My stronger guilt defeats my strong intent.
> . . . .
> In the corrupted currents of this world,
> Offence's gilded hand may shove by justice,
> And oft 'tis seen the wicked prize itself
> Buys out the law: but 'tis not so above;
> There is no shuffling, there the action lies

> In his true nature; and we ourselves compelled,
> Even to the teeth and forehead of our faults
> To give in evidence. What then? What rests?
> Try what repentance can: what can it not?
> Yet what can it when one cannot repent?
> O wretched state! O bosom black as death!
> O limèd soul, that, struggling to be free,
> Art more engaged! Help, angels! Make assay!
> Bow, stubborn knees!"
>
> *Hamlet*, act iii. sc. 3.

And they cannot! But the knees that cannot bend are in presence of the hosts of which Hugo speaks. The knees that cannot bend are dead. Is the deepest final laughter of the soul at itself a laughter from which it can flee? In the next life shall we escape these internal bursts of laughter from conscience? Not unless the soul can escape from itself. While we continue to be spiritual individualities, we must keep company with the plan of our natures; and this plan is expressed in that allegory of Romainville, lilacs in April, and the question from the half-headless host, "Do you not know that you have been dead a long time?"

There is in conscience, Bishop Butler says, a prophetic office; and it is to be regretted that the foremost Christian apologist of the late centuries did not develop this stupendous thought, which he only suggests in his famous sermons. "Conscience, without being consulted," Butler says, "magisterially exerts itself, and, if not forcibly stopped, naturally and always of course goes on to anticipate a higher and more effectual sentence, which shall hereafter second and affirm its own. But this part of the office of conscience," continues Butler, "is beyond my present design explicitly to consider." (*Upon Human Nature*, Ser. 11.) Now, precisely where Butler paused in his consideration of the prophetic office of conscience, Shakspeare seems to have begun : —

> " To be, or not to be, — that is the question.
> . . . .
> To die, to sleep ;
> To sleep ! perchance to dream ; ay, there's the rub.
> For in that sleep of death what dreams may come,
> When we have shuffled off this mortal coil,
> Must give us pause.
> . . . .
> The dread of something after death, —
> The undiscovered country, from whose bourn
> No traveller returns, — puzzles the will,
> And makes us rather bear those ills we have
> Than fly to others that we know not of.
> *Thus* conscience does make cowards of us all."
> *Hamlet*, act iii., sc. 1.

You say that Shakspeare is here speaking poetically? But again and again he utters the same thought. You remember Clarence's dream : —

> " My dream was lengthened after life.
> Oh ! then began the tempest to my soul,
> Who passed, methought, the melancholy flood,
> With that grim ferryman the poets write of,
> Unto the kingdom of perpetual night.
> The first that there did greet my stranger soul
> Was my great father-in-law, renownèd Warwick,
> Who cried aloud, 'What scourge for perjury

> Can this dark monarchy afford false Clarence?'
> And so he vanished; then came wandering by
> A shadow like an angel, with bright hair
> Dabbled in blood; and he squeaked out aloud, —
> ' Clarence is come, false, fleeting, perjured Clarence,
> That stabbed me in the field by Tewksbury.
> Seize on him, Furies! take him to your torments!'
> With that, methought, a legion of foul fiends
> Environed me about, and howled in mine ears
> Such hideous cries, that, with the very noise,
> I, trembling, waked, and for a season after
> Could not believe but that I was in hell."
>
> *King Richard III.*, act i. sc. 4.

The internal burst of laughter! Shakspeare knew what it was in its earlier smiles, or he could not have written these passages concerning souls that seem to have heard that laughter in its deepest final tones. [Applause.]

Out of the multitude of historical examples of the laughter of the soul at itself, take only two. There is Charles IX. of France. He consented to the massacre of St. Bartholomew. He is dying. He is twenty-four years of age. He is in such an agony of remorse that the historians say there is documentary evidence of the fact that he sweat blood. Not only did the blood pour out of nostrils and the corners of the eyes, but in many places through the corrugated veins did the blood ooze. That is history, and not poetry. He recalled the massacre of St. Bartholomew, to which he had assented. "How many murders! what rivers of blood!" and he went hence, as Clarence went out of his dream. "Quelle preuve," adds a French historian to his narrative of this scene (DURUY, *Histoire de France*, tome 2, p. 120), " de l'impuissance du crime à tromper la conscience du coupable!" You say that this is a very penetrating gleam into the recesses of natural law, if it be a fact. You know that facts of this kind are numerous in history; and no philosophy is sound that does not match itself to all the facts of its field. The blisses and pains of conscience! We know the pains better than the blisses; but the nature of things weighs as much for us as it does against us. The weight of the word ought is as great when it is against us, as it is when it is for us.

John Randolph fought a duel with Henry Clay. He walks into the senate-chamber, staggering in his last illness. Mr. Clay is rising to speak. The two men have not addressed each other for months. "Lift me up," says Randolph, loud enough for Clay to hear him: "I must listen to that voice once more." He was lifted up; Clay finished his speech; and the men shook hands, and parted almost friends. Randolph was taken to Philadelphia, and his biographer (*Life of Randolph*, vol. ii., last chapter) — I am citing no newspaper clamor — affirms that on his death-bed he asked his physician to show him the word remorse in the dictionary. "There is no dictionary in the room," says the physician. "Very well: here is a card. The name of John Randolph is on one side of it: write on the other the word which best symbolizes his soul. Write remorse in large letters; underscore the word." After that was done, Randolph lifted up the card before his eyes, and repeated in a loud voice, three times, "Remorse, remorse, remorse!"—"What shall we do with the card?" says the physician. "Put it in your pocket, and when I am dead look at it." You say Randolph was insane. After all these acts he dictated his will, manumitting his slaves; and at that day such a will could not be drawn, except by an acute and clear head. It was technically perfect. "You know nothing of remorse," said John Randolph, no theologian, no partisan, a man of the world. "I hope I have looked to Almighty God as a Saviour, and obtained some relief; but when I am dead look at the word which utters the inmost of my soul, and you will understand of what human nature is capable." He had heard the internal burst of laughter, although perhaps not in its deepest tones.

To summarize now what these examples prove:—

1. There is an Eternal Power, not ourselves, which makes for righteousness.

2. An entire agreement exists between conscience and the issues of things.

3. Our consciences are thus in harmony with that Power.

4. We are compelled to judge ourselves according to the moral ideals authorized by this Eternal Power, not ourselves, which makes for righteousness.

5. We cannot escape from this Power.

6. We must be in either harmony or dissonance with it.

7. If in dissonance with it, we must bear the pains

which are the inevitable penalties of such dissonance.

8. Conscience thus makes cowards of us all.

9. It does so not only by the fear of moral penalty in this life, but by the fear of something after death.

10. The constitutional fear of "something after death," of which Shakspeare and Butler speak, is a proof that there is something there.

11. While the prophetic action of conscience thus intensifies all the pains of conscience, it may also intensify all its blisses.

12. It is true, on the one hand, that the innermost laughter of the soul at itself, it rarely hears more than once without hearing it forever.

13. It is true, on the other, that the innermost benediction of the soul upon itself, it rarely hears more than once without hearing it forever.

14. The innermost laughter and the innermost benediction come from the depth of conscience.

15. But the weight of the word ought is a revelation of the nature of things.

16. The nature of things is only another name for the Divine Nature.

17. The laughter of the soul and the benediction of the soul as to itself, in the innermost of conscience, are the laughter and benediction of the nature of things; that is, the benediction and the laughter of the Lord. [Applause.]

18. The laughter of the soul at itself is a laughter from which it cannot flee.

X.

SHAKSPEARE ON CONSCIENCE.

THE NINETIETH LECTURE IN THE BOSTON MONDAY LECTURE-
SHIP, DELIVERED IN TREMONT TEMPLE, DEC. 3.

He who resolveth to do every duty, is immediately conscious of the presence of the gods. — BACON.

> Stern lawgiver! yet thou dost wear
>   The Godhead's most benignant grace;
> Nor know we any thing so fair
>   As is the smile upon thy face;
> Flowers laugh before thee on their beds,
> And fragrance in thy footing treads;
> Thou dost preserve the stars from wrong;
> And the most ancient heavens through thee are fresh and strong.
>   WORDSWORTH: *Ode to Duty*.

X.

SHAKSPEARE ON CONSCIENCE.

PRELUDE ON CURRENT EVENTS.

SOMETIMES in ancient Athens, previous to elections, the streets were swept with a vermilion-colored cord. In the assemblies which Demosthenes addressed at the Pynx, no important law could be passed unless six thousand votes in its favor were deposited in the urns. To secure an audience of the necessary size, servants of the state were sent through the market-place with a rope chalked red; and whoever received a stain on his toga, as that never-loitering line, stretched from side to side of the streets, passed along the crowded ways, was fined as an enemy of the state. Charles Sumner often affirmed that the citizen who neglects his political duties is a public enemy. A law of Pythagoras pronounced every free man infamous who in questions of public moment did not take sides. Compulsory voting was the rule in ancient Athens; and one could almost wish that it were in modern America. We should be imitating the Athenians if we were to double the poll-tax of all who can vote and do not.

Athenian scholars like President Seelye and President Chadbourne are not lowering their dignity at all by endeavoring to teach us through their personal example the mission of the scholar in politics. [Applause.] They take no partisan stand, but simply a patriotic one. Assuredly terror would blanch the cheeks of political corruption if such examples could be followed as widely as they are already honored. Our fathers taught, and so have an hundred years of American history, that eternal vigilance, and not merely endless grumbling and sour grimace on the part of culture, is the price of liberty. How utterly has the mood of scholarly patriotism changed in the last fifty or eighty years! A citizen of Brooklyn, not long ago, said to me that he supposed that of course the mayor of Boston speaks Latin. In the old days we were jealous of our rights on these three hills. But now that a foreign-born population has taken an honored place at the ballot-box, — a position from which we do not wish to drive them at all, — some of us are too lofty to ask any favors of them unless they first will ask favors of us by putting our names in nomination. If the foreign-born vote, if half-educated suffrage, if that part of our population which Lord Beaconsfield would say is unfit in many particulars for citizenship, will ask our permission and come to us and burn sufficient incense, perhaps we can go down to the hustings and the ballot-box, and attend to our duties as American citizens, attracted thither by the smell of praise. If culture cannot rule, then culture will secede from politics.

Class secession is hardly less dangerous than geographical secession. The withdrawal of the cultivated class from politics may ultimately work as much harm as the open secession of our Carolinas and Georgias from the Union. [Applause.] Class secession from politics is often actuated by much the same thoughts as those which governed the geographical secession of States. We are not believers in democracy. We desire to have our rights respected without defending them. We, as an educated and propertied class, are by and by in a new organization of society in America, to have the privileges that belong of natural right to culture and wealth. There will be a new order of government brought into existence ultimately on this continent. Give America two hundred people to the square mile, and count of heads and clack of tongues will not keep life and property safe here. Democracy is failing. We will not be in at its death. We will wait until our great cities have suffered enough to put their interests into the hands of wealth, and to insist on a property qualification for the franchise. Already the whisper of fleeced municipal tax-payers grows loud in many commercial circles, that a stronger government is needed in America than ever the many can exercise over the many.

For one, I believe the young men of the United States have as a mass given up even the unexpressed fear that we shall abandon Democracy. These feelings of some of the cultured, that new arrangements will be made, are not shared by the remnant of the generation which preserved the Union. A large part of the young men of America who should now be entering on patriotic public careers are already in their graves. In this country, my generation is a fragment. It is a tattered remnant left over after battle. We have already laid down many lives, that men may have the right of franchise. We have done something for the unification of this country. We are willing to do more. May the right arms of the young men of America drop from their sockets, and may their tongues cleave to the roofs of their mouths, if they ever forget that their brethren died, not only for the unification, but for the purification of this nation [applause], or if they ever fail to endeavor in politics, in social life, in the pulpit, on the platform, in the press, to sell their lives as dearly in the purification of America, as their brethren have sold theirs in its unification. [Applause.]

I thank Providence, that the young citizens' political committees are acting as if they believe that Democracy must try out its own problems, and must purify the ballot-box to begin with. We have committees organized in this city in one of the parties, and I wish they were organized in both, to clarify registration-lists; to watch ballot-boxes; to see to it that the press is prompted occasionally in the proper direction; and, above all, to inspirit public sentiment, by throwing the power of the parlor and the platform and the pulpit at the right moments toward the just side, when make-weights are of commanding consequence in closely-contested elections. There are in this city no peculiar corruptions. Undoubtedly Boston politics are better than those of New York or Chicago, on the whole; but that would not be saying any thing greatly to our praise. There are, however, in this city, young men's committees on the watch; and it turns out, as a practical result, that an immigrant cannot be made a voter now unless he is personally present before the recording officers to take the oath and to sign the declaration that he becomes a citizen. There are young men's committees on the watch; and every ballot-box in this city will be managed according to law, if the young men have their way; and if they do not have their way, and the boxes are not thus managed, the young men, aided by persons older than they, — wisdom with the aged, action with the young, — mean to prosecute every case of violation of law to conviction. [Applause.] There have been many shrewd arrangements made here, as elsewhere, for the violation of law at the ballot-boxes; but either these arrangements will be defeated, or somebody will suffer if they are carried out.

In one of the largest cities of the West an election committee announced, not long ago, that the day has gone by when it can be expected that the cultivated class in our great towns will take any active part in politics. A friend of mine, who was in a pulpit in that city, but who now, thank God! is a minister in New England, and who never preached politics in the pulpit, went down to the ballot-boxes of a corrupt ward, and challenged votes on several different occasions, and did so all alone. The opposing party put three or four roughs near him. Although they did not attack him physically, they filled the day with profanity and obscenity, and endeavored to drive away all decent men by their Harpy clamor. The scholar held his place, threatened prosecution against lax officers behind the ballot-boxes, and the result was that a dark ward was illuminated, if not by noon, at least by twilight, and many a wild beast of politics ran to his den. [Applause.] From that single example it became, in several wards of that city, the fashion for cultivated men to go down and challenge roughs at the polls. Many Englishmen like to do this.

John Bright says that he will not vote for a wider extension of the franchise in Great Britain, than is in existence there, because political absenteeism ruins a good cause every now and then. If we could have political absenteeism throttled, one feels almost sure that this sea of unrest in which many of us swim, this feeling that the universal franchise of America is to ruin her, would subside at last. The secession of culture from primary political meetings, and from the post of Argus at the ballot-box, is to be judged by its fruits. We are to assert our rights, or have none. We are to occupy our privileges, or find them, little by little, curtailed. Even if you could do but little at the polls, you might do much at the primary meetings. Even if you could do little in the latter places, you might do much by inspiriting young men who have time for the work, to attend to the patriotic duties of unmasking fraud.

Lord Bacon affirms that the best materials for political prophecy are the unforced opinions of young men. In this Commonwealth, when Charles Sumner and Henry Wilson were beginning their career, the use of that Baconian method of forecast might have been profitable to both the timid friends and the haughty opponents of just reform. If the young men of America enforce the suffrage laws, they will have the sound part of the press of every political party on their side. Let them use their opportunities resolutely, and politics — which are only a weather-vane — will show which way the wind blows. The American people are Æolus' cave. If in the national sea the political ship rots in calms, or sails in the wrong direction, the fault is not as much with the pirates on board as with Æolus, who might awaken hurricanes in his mountain, and send them forth to make Æneas pray as he never did of old when he had lately left Dido, or when the jealous Juno shook the Trojan fleet.

THE LECTURE.

Whom does Shakspeare make us admire? An author is what he causes us to love. Do we find ourselves retaining to the end our respect for Falstaff? Henry V., who had toyed with vice in Falstaff's company, rejects the gray-haired lecher after becoming king.

"*The King* to *Falstaff.* I know thee not, old man; fall to thy
prayers;
How ill white hairs become a fool and jester!
I have long dreamed of such a kind of man,
So surfeit-swelled, so old and so profane;
But being awaked, I do despise my dream.
Make less thy body hence, and more thy grace.
. . . .
Reply not to me with a fool-born jest;
Presume not that I am the thing I was,
For Heaven doth know, so shall the world perceive,
That I have turned away my former self;
So will I those that kept me company.
. . . .
I banish thee on pain of death,
As I have done the rest of my misleaders,
Not to come near our person by ten mile."
2 *King Henry IV.*, act v. sc. 5.

Although Falstaff is pictured in detail, Shakspeare plainly intends that we shall not permanently admire him. In the end he crushes even our animal regard for Sir John by making him die a loathsome death. "Let thy blood be thy direction till thy death!" says Shakspeare: "then if she that lays thee out says thou art a fair corse, I'll be sworn, and sworn upon it, she never shrouded any but lazars." (*Troilus and Cressida*, act ii. sc. 3.) Do we love Iago? Shakspeare pictures him, too, in great detail; but on the whole our feeling in his presence is that which comes to us when we look into a serpent's eye.

There are roisterers and feather-heads reflected in the lower half of Shakspeare's mirror; but if you will fathom your own experience with this writer, you will find that it is not the lower, but the upper half of his far-spread and astoundingly faithful glass, that captures you permanently. I am not, perhaps, advanced enough in life to understand Shakspeare; it is said that no man under forty can read Shakspeare; but, as I grow older, I am more and more attracted to the upper half, or, I may say, to the upper quarter, of his mirror. He holds up the picturing glass to all that is; and undoubtedly, in a full representation of human nature, especially as it was forced on Shakspeare's attention in a roistering court and in the life of the London of the days of Queen Elizabeth, there will be blotches in the lower half of the reflecting glass. But the final impression Shakspeare seems to make is that the upper half of the mirror was himself. He dwells in his advanced years more upon the Unseen, upon the moral law, upon the great characters of his tragedies, and less and less, except as a foil, upon the lower traits and the coarser in human nature.

Indeed, if I were to select out of all Shakspeare's characters the one person whom he most resembles, I should take Henry V. That soul was equipped for peace or war, for sport or earnest, for the light things of the day of harmless play, or the stern things of loud resounding contest. And he grew better, Henry V. did, as he grew older. It is true, he had been a companion of Falstaff; no doubt his youth had many things in it which he deserved to regret; but he grows as his years advance, and when kingship comes to him, he is a hero, one of the most full-orbed of all the characters delineated on Shakspeare's canvas. Hamlet? He was like Shakspeare in several very great things, but he did not love action. He was almost insanely dilatory in cases of the highest importance; but Shakspeare had decision as well as gentleness. A not unsuccessful practical activity, we know, filled a considerable part of his life. For the benefit of a softer and less strenuous age than his own, and almost as if the false standards of the school of *Genialität* in literature were foreseen by him, he drew in Hamlet, I think, a balanced criticism of high intellectual power and subtle intensities of emotion not conjoined with sufficient executive capacity.

Shakspeare knew better and better, as he grew older, what Kant affirmed in his last years, that the best melody of the harp never is obtained until the chords are stretched tightly, and the plectrum with which the resonant wires are struck is made firm. Madame de Staël says of Schiller, that his Muse was Conscience. His poetry has several of its high crystalline fountain-springs in the heights of Kant's philosophy. But even Schiller once complained that Kant's system of ethics occasionally takes on the aspect of a repulsive, hard, imperative morality, and is not attractive. Kant replied that the two objects of moral training are to give "hardihood" in the application of conscience to the motives, and "gladness" in prompt and full obedience to the moral sense. (*Metaphysics of Ethics*, edition by Semple.)

Hardihood! That is the stretching of the chord tautly in the harp. Hardihood! That is the firmness of the plectrum which smites the chord. Hardihood! That is the first object of moral training. Gladness is the second, but that is only the music derived from the tightly-stretched chord and the firm plectrum. More and more, as Shakspeare grew older, he tightened the moral strands in the colossally wide harp of his nature, and the stretched chords he struck with firm plectra, and their far-resonant upper notes at last are harmonious with the deep bass of the moral law in the nature of things. That is Shakspeare. [Applause.] Here is the last tone shed from Shakspeare's harp within the hearing of this world: "I commend my soul into the hands of God, my Creator, hoping, and assuredly believing, through the only merits of Jesus Christ my Saviour, to be made partaker of life everlasting." (*Shakspeare's Will.*)

Undoubtedly he was an American in his youth. He thought that good music could be produced by leaving the chords in delightfully uncertain positions. A firm plectrum! Why, no; it would not be liberal to make the plectrum solid! It would not be in harmony with advanced thought to tighten the chords! Hardihood! Why, the very word is odious to luxurious liberalism! Hardihood! Schiller protests against Kant, when he misunderstands him, not knowing that hardihood is the mother of gladness in the harp.

Shakspeare in his youth, no doubt, married too

early, and yet none too early; and to this keen, self-imposed curse he has himself again and again made allusion. I beg your pardon; you must meditate in secret over these stains of blood in Shakspeare's writings. Do you remember that he says that on certain conditions heaven will bless a marriage, and on certain other conditions will not? Perhaps Henry V. did not perceive the kingship that was before him. Undoubtedly Shakspeare, who for a hundred years after his death was not widely worshipped, did not understand what kingship was awaiting him. As Henry V. strengthened himself the moment he became king, so Shakspeare would have done if he could have seen in advance the enduring responsibilities of the regnancy which literature was providing for him. But, had he foreseen this, he could not have tightened more strenuously than he did one chord in his harp.

If the fact, without the form, of marriage, exists before

> "All sanctimonious ceremonies may
> With full and holy rite be ministered,
> No sweet aspersion shall the heavens let fall
> To make this contract grow; but barren hate,
> Sour-eyed disdain and discord, shall bestrew
> The union of your bed with weeds so loathly
> That you shall hate it both; therefore take heed
> As Hymen's lamps shall light you."
>
> *Tempest*, act iv. sc. 1.

See Winter's Tale, act i., line 278, "before her troth-plight." Also, White's Shakspeare, vol. i. pp. xxix.-xxxvi.

Shakspeare did not know through how many hundreds of years these words would be read over his tomb in Stratford-on-Avon, and how many times they would recall the crime of a woman eight years older than he, and his own infamy; but he would not have erased them, could he have foreseen all.

When men in our day strike the lower chords of their nature loosely; when we are taught by advanced materialists that we are not responsible, whatever we do; when Häckel asserts that the will is never free; when a professor, possessed of excellent literary capacity, and reverenced throughout the civilized world as a leader in physical science, stands up and maintains, as Tyndall did at Birmingham lately, in so many words, that "the robber, the ravisher, and the murderer offend because they cannot help offending," then I like to look across that green shield, sir [turning to the Rev. Mr. Rainsford], called England, circled by the azure deep, and to remember that Birmingham and Stratford-on-Avon lie not far apart, as bosses on that buckler of the world's good sense. Lord Bacon said that he wished a science of the human passions could be elaborated. Gervinus, the best German commentator on Shakspeare, affirmed that if Bacon had turned to his neighbor William he might have had such a science, and that one to-day might be constructed from his works. Tyndall stands at Birmingham, and proclaims, as Häckel has taught, that the science of the human passions must include the assertion that the will is never free. Lord Bacon, I think, feels uneasy on his pedestal at such science. At any rate, Gervinus on the Rhine, in his tomb, whispers yet to civilization, that William Shakspeare,

Bacon's contemporary, will teach us the true theory of the passions. When Tyndall utters at Birmingham his famous assertion that the robber, the ravisher, the murderer, offend because they cannot help offending, I turn to this grave at Stratford-on-Avon, — the grave of an honest man; for we have seen the epitaph its occupant has put upon himself, and how little he excused any of his own misdeeds, — and I listen. I hear words, three hundred years old, indeed; but I recommend them, in spite of their antiquity, as a motto for Tyndall's address: —

"This is the excellent foppery of the world, that, when we are sick in fortune, — often the surfeit of our own behavior, — we make guilty of our disasters the sun, the moon and stars; as if we were villains by necessity." [Applause.] Professor Tyndall hears that at Birmingham? "Fools by heavenly compulsion; knaves, thieves, and treachers, by spherical predominance; drunkards, liars, and adulterers, by an enforced obedience of planetary influence; and all that we are evil in, by a divine thrusting on." Does Tyndall listen? "An admirable evasion of abominable man, to lay his goatish disposition to the charge of a star! My nativity was under Ursa Major; so that it follows that I am rough. Tut! I should have been that I am, had the maidenliest star in the firmanent twinkled on my birth." [Applause.] (*King Lear*, act i. sc. 2.)

But it is impossible to condense a tithe of what ought to be said concerning Shakspeare's views on conscience, into the hand's-breadth of time allowed me here. Let me notice the leading questions to which he gives answers, although I cannot recite all the replies.

1. Whom does Shakspeare make us admire?
2. Whom does he make responsible for sin?
3. Does Shakspeare make the word *ought* heavier than any other syllable?
4. Does Shakspeare teach that there is a God in conscience?
5. Does he give conscience a prophetic office?
6. Does Shakspeare make judicial blindness one of the inevitable penalties of the suppression of light?
7. May conscience, in the opinion of Shakspeare, make cowards of us all?
8. How, according to this poet, does conscience color the external world?
9. Does Shakspeare admit that conscience may cease to exist or to act in the incorrigibly evil?
10. What, according to Shakspeare, are some of the physical effects of conscience?
11. Does he teach that conscience may produce despair?
12. Is Shakspeare supported in his conclusions by other poets?

As one would touch the multiplex array of banks of organ-keys at random to test the tones of some mighty instrument, so I open a copy of Shakspeare at random, with no partisan plea to make. What massive and overawing tones are these first ones I happen to strike! —

"In the great hand of God I stand."

Why? Because I am following my conscience in opposing a bloody tyrant.

> "And thence
> Against the undivulged pretence I fight
> Of treasonous malice."
> *Macbeth*, act ii. sc. 3.

But here is a contrasted tone strangely deep: —

> "What do I fear? myself? There's none else by;
> Richard loves Richard; that is, I am I.
> Is there a murderer here? No; yes, I am;
> Then fly. What, from myself? Great reason; why?
> Lest I revenge. What? Myself upon myself?
> Alack! I love myself. Wherefore? for any good
> That I myself have done unto myself?
> Oh, no! alas! I rather hate myself,
> For hateful deeds committed by myself."
> *King Richard III.*, act v. sc. 3.

> "The weariest and most loathèd worldly life
> That age, ache, penury, and imprisonment
> Can lay on nature, is a paradise
> To what we fear of death."
> *Measure for Measure*, act iii. sc. 1.

> "The dread of something after death
> ... puzzles the will.
> Thus conscience does make cowards of us all."
> *Hamlet*, act iii. sc. 1.

> "Conscience is a thousand swords."
> *King Richard III.*, act v. sc. 2.

Strike the peaceful, cheering, mysteriously commanding notes once more: —

"What stronger breastplate than a heart untainted?
Thrice is he armed that hath his quarrel just,
And he but naked, though locked up in steel,
Whose conscience with injustice is corrupted."
2 *King Henry VI.*, act iii. sc. 2.

"Be just, and fear not.
Let all the *ends thou aim'st at* be thy country's,
Thy God's, and truth's; then, if thou fall'st, O Cromwell,
Thou fall'st a blessed martyr."
King Henry VIII., act iii. sc. 2.

"Now, for our consciences, the arms are fair,
When the *intent* of bearing them is just."
King Henry IV. sc. 3.

"My wooing mind shall be expressed
In russet yeas and honest kersey noes."
Love's Labor Lost, act v. sc. 2.

"That which you speak is in your conscience washed."
King Henry V., act i. sc. 2.

"What motive may
Be stronger with thee than the name of wife?
That which upholdeth him that thee upholds, —
His honor: oh, thine honor, Lewis, thine honor."
King John, act iii. sc. 1.

"A peace above all earthly dignities,
A still and quiet conscience."
Henry VIII., act iii. sc. 2.

Strike the contrasted notes again: —

"*First Murderer.* — So when he opens his purse to give us our reward, thy conscience flies out.

Second Murderer. — Let it go; there's few or none will entertain it.

First Murderer. — *How if it come to thee again?*

Second Murderer. — I'll not meddle with it. It is a dangerous thing. It makes a man a coward. A man cannot steal, but it accuseth him; he cannot swear, but it checks him; tis a blushing, shamefaced spirit that mutinies in a man's bosom; it fills one full of obstacles; it made me once restore a purse of gold that I found; it beggars any man that keeps it; it is turned out of all towns and cities for a dangerous thing.

First Murderer. — Zounds, it is even now at my elbow."

King Richard III., act i. sc. 4.

"My conscience, hanging about the neck of my heart, says very wisely to me, 'Budge not.' — 'Budge,' says the fiend. 'Budge not,' says my conscience."

Merchant of Venice, act ii. sc. 2.

"I, I myself, sometimes, leaving the fear of God on the left hand and hiding mine honor in my necessity, am fain to shuffle, to hedge, and to lurch."

Merry Wives of Windsor, act ii. sc. 2.

"Put up thy sword, traitor,
Who mak'st a show, but durst not strike, thy conscience
Is so possessed with guilt. Come from thy ward,
For I can here disarm thee with this stick,
And make thy weapon drop."

Tempest, act i. sc. 2.

"O heaven, put in every honest hand a whip
To lash the rascals naked through the world."

Othello, act iv. sc. 2.

"The color of the king doth come and go
Betwixt his purpose and his conscience
Like heralds 'twixt two dreadful battles set.
His passion is so ripe, it needs must break."

King John, act iv. sc. 2.

"The grand conspirator,
With clog of conscience and sour melancholy
Hath yielded up his body to the grave. . . .
The guilt of conscience take thou for thy labor,
With Cain go wander through the shades of night."

King Richard II., act v. sc. 2

"The worm of conscience still begnaw thy soul."
King Richard III., act i. sc. 3.

I open the book again, and hear Shakspeare answer the question whether blindness sent as a judgment results from the suppression of light. Lady Macbeth says,—

> "The raven himself is hoarse
> That croaks the fatal entrance of Duncan
> Under my battlements. Come, you spirits
> That tend on mortal thoughts, unsex me here,
> And fill me from the crown to the toe top-full
> Of direst cruelty! make thick my blood;
> Stop up the access and passage to remorse,
> That no compunctious visitings of nature
> Shake my fell purpose, nor keep peace between
> The effect and it! Come to my woman's breasts,
> And take my milk for gall, you murdering ministers,
> Wherever in your sightless substances
> You wait on nature's mischief! Come, thick night,
> And pall thee in the dunnest smoke of hell,
> That my keen knife see not the wound it makes,
> Nor heaven peep through the blanket of the dark
> To cry, 'Hold, hold!'"
> *Macbeth*, act i. sc. 4.

The prayer was answered. Never, since it was written in the Bhagvat Gheeta that "repeated sin impairs the judgment," and that "he whose judgment is impaired sins repeatedly;" never, since the Spanish proverb was invented that "every man is the son of his own deeds,"— has the law of judicial blindness been proclaimed with such sublimity as in this utterly unpartisan and secular passage. Macbeth himself, under similar circumstances, says:—

> . . . . . "Come, seeling night,
> Cancel and tear to pieces that great bond
> Which keeps me pale! Light thickens; and the crow
> Makes wing to the rooky wood."
> *Macbeth*, act iii. sc. 2.

A fiend in human form in Titus Andronicus has made evil his good: —

> "*Lucius.* — Set him breast-deep in earth, and famish him,
> There let him stand and rave, and cry for food.
> *Aaron.* — I am no baby, I, that with base prayers
> I should repent the evils I have done;
> Ten thousand worse than ever yet I did
> Would I perform if I might have my will.
> If one good deed in all my life I did,
> I do repent it from my very soul."
> *Titus Andronicus*, act v. sc. 3.

Elsewhere Shakspeare affirms most definitely that it is a pervasive natural law that, —

> "When we in our viciousness grow hard,
>   (O misery on't!) the wise gods seel our eyes;
> In our own filth drop our clear judgments, make us
> Adore our errors; laugh at us, while we strut
>   To our confusion."
> *Antony and Cleopatra*, act iii. sc. 13.

Is there a God in conscience?

> "Methinks in thee some blessed spirit doth speak
>   His powerful sound within an organ weak."
> *All's Well That Ends Well*, act ii. sc. 1.

> "I hold you as a thing ensky'd and sainted,
>   By your renouncement an immortal spirit,
>   And to be talked with in sincerity,
>   As with a saint!"
> *Measure for Measure*, act i., sc. 3.

When Shakspeare is called on to paint despair, he makes the elements themselves draw the picture.

> " Oh, it is monstrous, monstrous!
> Methought the billows spoke and told me of it:
> The winds did sing it to me, and the thunder,
> That deep and dreadful organ-pipe, pronounced
> The name of Prosper: it did bass my trespass."
> *Tempest*, act iii. sc. 3.

You know Arthur was about to be murdered, and that Hubert was suspected of the murder; and when there is a confronting of that crime with the light of conscience, Shakspeare makes one of his characters say, —

> " Beyond the infinite and boundless reach
> Of mercy, if thou didst this deed of death,
> Art thou damned, Hubert."

Really, I beg pardon for reading this in Boston, and so near Indian Orchard. [Laughter.]

> " If thou didst but consent
> To this most cruel act, do but despair;
> And if thou want'st a cord, the smallest thread
> That ever spider twisted from her womb
> Will serve to strangle thee; a rush will be a beam
> To hang thee on; or, wouldst thou drown thyself,
> Put but a little water in a spoon,
> And it shall be as all the ocean,
> Enough to stifle such a villain up."
> *King John*, act iv. sc. i.

This serious observer represents ruin as possible to man: —

"Oh, she is fallen
Into a pit of ink, that the wide sea
Hath drops too few to wash her clean again;
And salt too little which may season give
To her foul-tainted flesh."
Much Ado About Nothing, act iv. sc. i.

Shakspeare is nowhere a partisan. He lived between two conflicting currents, men that were sometimes called fanatics, but who have founded New England, — quite a piece of work in the world, — and the rough, roistering circles of the court. Shakspeare was no fanatic, but he was no roisterer. In few words, he sums up, in a passage more terrific, probably, than any other in his dramas, the whole effect, mental and physical, of an upbraiding conscience. How does this man, speaking to roisterers in his own audience, and writing under the fear that he was to be called illiberal, and sneered at for sympathizing with fanatics, — how does this man, to whom human nature lay open, draw the picture of a soul accusing itself?

"*Macbeth.* One cried, 'God bless us!' and 'Amen,' the other:
As they had seen me with these hangman's hands,
Listening their fear, I could not say 'Amen,'
When they did say, 'God bless us.'"

What are the physical effects of an outraged moral sense? Shakspeare has answered.

"*Lady Macbeth.* Consider it not so deeply.
Macbeth. But wherefore could not I pronounce 'Amen'?
I had most need of blessing, and 'Amen'

Stuck in my throat.

Lady Macbeth. These deeds must not be thought
After these ways: so it will make us mad.

Macbeth. Methought I heard a voice cry, 'Sleep no more!
Macbeth does murder sleep;' the innocent sleep:
Sleep, that knits up the ravelled sleeve of care,
The death of each day's life, sore labor's bath,
Balm of hurt minds, great nature's second course;
Chief nourisher in life's feast.

Lady Macbeth. What do you mean?

Macbeth. Still it cried, 'Sleep no more!' to all the house:
'Glamis hath murdered sleep, and therefore Cawdor
Shall sleep no more; Macbeth shall sleep no more.'

Lady Macbeth. Who was it that thus cried? Why, worthy thane,
You do unbend your noble strength, to think
So brain-sickly of things. Go get some water,
And wash this filthy witness from your hand.
Why did you bring these daggers from the place?
They must be there. Go carry them; and smear
The sleepy grooms with blood.

Macbeth. I'll go no more;
I am afraid to think what I have done:
Look on't again, I dare not.

Lady Macbeth. Infirm of purpose!
Give me the daggers: the sleeping and the dead
Are but as pictures; 'tis the eye of childhood
That fears a painted devil. If he do bleed
I'll gild the faces of the grooms withal;
For it must seem their guilt.

Exit. Knocking within.

Macbeth. Whence is that knocking?
How is't with me when every noise appalls me?
What hands are here? Ha! they pluck out my eyes.
Will all great Neptune's ocean wash this blood
Clean from my hand? No; this my hand will rather
The multitudinous seas incarnadine,
Making the green one red." *Macbeth,* act ii. sc. 2.

But if Macbeth had read Professor Tyndall's speech at Birmingham, undoubtedly advanced thought would have washed the murderer's red right hand.

To summarize all that Shakspeare has said, therefore, take this opinion from Gervinus: —

"The deity in our bosoms Shakspeare has bestowed with intentional distinctness, even upon his most abandoned villains, and that, too, when they deny it. To nourish this spark, and not to quench it, is the loud sermon of all his works." (GERVINUS, *Commentaries on Shakspeare*, p. 910.)

Do you say that, after all, Shakspeare was morbid on a few points? Well, if he was, Lord Byron was not. Omitting Milton, Schiller, and Dante, and Euripides, Sophocles, and Æschylus, who, on the subject of Conscience, agree with Shakespeare only too startlingly, we will take Byron as a fair answer to the question, whether other poets sustain the prophet and philosopher of Stratford-on-Avon. Lord Byron had guilt of which he knew the extent, and which God has not suffered to be known to men at large, and I hope never will suffer to be known. But this poet, understanding very well that the world was listening, and that on every sentence of his concerning the moral sense and remorse a microscope would be placed age after age, does not hesitate to say, —

> "Yet still there whispers the small voice within,
> Heard through God's silence, and o'er glory's din;
> Whatever creed be taught, or land be trod,
> Man's conscience is the oracle of God."
>
> BYRON, *Island*.

> "But at sixteen the conscience rarely gnaws
> So much as when we call our old debts in
> At sixty years, and draw the accounts of evil,
> And find a deuced balance with the Devil."
>
> BYRON.

Here are the most incisive and perhaps the most self-revelatory words Byron ever wrote concerning Conscience: —

> "The mind that broods o'er guilty woes
> Is like the scorpion girt by fire :
> In circle narrowing as it glows,
> The flames around their captive close ;
> Till inly scorched by thousand throes,
> And inly maddening in her ire,
> One and sole relief she knows, —
> The sting she nourished for her foes,
> Whose venom never yet was vain,
> Gives but one pang, and cures all pain,
> She darts into her desperate brain.
> So do the dark in soul expire,
> Or live like scorpion girt by fire ;
> So writhes the mind remorse hath riven,
> Unfit for earth, undoomed for heaven ;
> Darkness above, despair beneath,
> Around it flame, within it death."
>
> BYRON, *Giaour*.

[Applause.]

THE INDEPENDENT.

THE LARGEST, THE ABLEST, THE BEST, RELIGIOUS NEWSPAPER IN THE WORLD.

It will maintain the high standard of excellence which has characterized it for the first thirty years of its history. It pays more money for contributed articles from the best writers and thinkers of the country than any other three religious newspapers. Its Editorial Department will treat upon the Religious, Political, and Social questions of the day, in a frank, free, and fearless manner.

REV. JOSEPH COOK'S LECTURES

In Tremont Temple, Boston, which have been published, *verbatim*, in THE INDEPENDENT during the past two winters, will be continued the present winter, being published *in full* with the preludes.

THE YALE LECTURES ON PREACHING

Are reported, verbatim, each year for THE INDEPENDENT, and should be read by every clergyman and layman in the land.

SERMONS

By eminent clergymen of all denominations will continue to be printed regularly each week.

THE DEPARTMENTS

Embrace Editorial, Editorial Notes, Religious News, Sunday-School, Literary, and Reviews of New Books, Science, Biblical Research, Fine Arts, Educational, Agricultural, Commercial, Financial, and Insurance. They are contributed to and edited by the best talent to be procured.

PREMIUMS.

We offer Rev. Joseph Cook's valuable new volumes, entitled, "Biology," "Transcendentalism," "Orthodoxy," "Conscience," "Heredity," and "Marriage," embodying in a revised and corrected form the author's previous remarkable Monday lectures. They are published in handsome book form by Houghton, Osgood & Co., of Boston.

We will mail a copy of either volume, post-paid, to any subscriber to the INDEPENDENT who remits us $3 for a year in advance; or any subscriber may remit $5.50, and we will send him the INDEPENDENT for two years in advance, and two volumes, post-paid; or, any three volumes, post-paid, to any one subscriber who remits $8.00 for three years in advance.

WORCESTER'S UNABRIDGED PICTORIAL QUARTO DICTIONARY.

Bound in sheep. 1,854 pages. Over 1,000 Illustrations. Issue of 1878.

We have made a special contract with the great publishing house of J. B. Lippincott & Co., of Philadelphia, by which we are enabled to offer the most desirable premium ever given by us or any other newspaper in the country. We will send this *Dictionary* to any person who will send us the names of *Three New Subscribers and Nine Dollars*; or who will, on renewing his own subscription, in advance, send us *Two New Names* additional and $9.00; or who will renew his own subscription for three years, in advance, and send us $9.00; or, for a new subscriber for three years and $9.00.

The regular price of the *Dictionary* alone at all the bookstores is $10.00, while the lowest price of three subscriptions is $9.00. Both the *Dictionary and the three subscriptions*, under this extraordinary offer, can therefore be had *together* for only $9.00. The *Dictionary* will be delivered at our office, or in Philadelphia, free, or be sent by express or otherwise from Philadelphia, as may be ordered, at the expense of the subscriber. The subscriber under this offer will not be entitled to any other Premiums.

Subscription Price, $3.00 per annum in advance, including any one of the following Premiums:—

Any one volume of the HOUSEHOLD EDITION OF CHARLES DICKENS'S WORKS, bound in cloth, with 16 Illustrations each, by Sol Eytinge.

MOODY AND SANKEY'S GOSPEL HYMNS AND SACRED SONGS, No. 2.

LINCOLN AND HIS CABINET; or, First Reading of the Emancipation Proclamation. Fine large Steel Engraving. By Ritchie. Size, 26 x 36.

AUTHORS OF THE UNITED STATES. Fine large Steel Engraving. 44 Portraits. Size, 24 x 38¼. By Ritchie.

CHARLES SUMNER. Fine Steel Engraving. By Ritchie.

GRANT OR WILSON. Fine Steel Engravings. By Ritchie.

EDWIN M. STANTON. Fine Steel Engraving. By Ritchie.

THE INNER LIFE OF ABRAHAM LINCOLN. By Frank B. Carpenter. Bound in cloth. 360 pages. It gives a better insight into his "inner life" than can be found elsewhere, and is altogether one of the most fascinating, instructive, and useful books of the kind ever published.

We offer one premium only for one year's subscription.

SUBSCRIPTION PRICE, $3.00 PER ANNUM, IN ADVANCE.

Specimen copies sent free. Address,

THE INDEPENDENT,

P. O. Box, 2,787.   NEW-YORK CITY.

BOSTON MONDAY LECTURES.

By JOSEPH COOK.

HISTORY OF THE LECTURES.

The Bibliotheca Sacra, January, 1878.

MR. JOSEPH COOK was invited, early in September, 1875, by the Young Men's Christian Association of Boston, to lead the noon prayer-meeting in the Meionaon daily for a week, and to make on each occasion an address of half an hour in length. After four of these services, it was found that the audience had quadrupled in size. Mr. Cook was requested to continue his addresses daily through another week. On Monday noon, Sept. 23, the subject was "Final Permanence of Moral Character; or, The Doctrine of Future Punishment," and it was noticed that a hundred ministers were in the audience. Mr. Cook was then requested to speak on the Atonement, on a Sabbath evening, in Park-street Church. He complied with this request, and spoke to an audience filling the house to its utmost capacity. He was then invited by the Young Men's Christian Association to speak every Monday noon, in the Meionaon, for twelve weeks. Oct. 25 his subject was "Boston Sceptical Cliques." "The Daily Advertiser" had a reporter present, who reproduced a part of the address. "The Springfield Republican" began to call attention to the large number of ministers and scholars who were present at the Monday Lectures. It was suggested, in many quarters, that these lectures should be continued regularly through the winter. Meantime, Mr. Cook was delivering one course of lectures at Amherst College, and another at Mount Holyoke Seminary, largely on Materialism, Evolution, and various biological topics. The Meionaon Hall seats about eight hundred persons, and in January, 1876, was completely filled by Mr. Cook's hearers. After four months had passed, the assemblies were occasionally gathered in Bromfield-street Church. The lectures continued to be under the auspices of the Young Men's Christian Association, until May, 1876, when, at a meeting in Bromfield-street Church, resolutions were passed founding the Boston Monday Lectureship, and placing it, for the next season, under the care of a committee, consisting of Prof. E. P. Gould of the Newton Theological Institute, the Rev. Dr. E. B. Webb of Boston, the Rev. Dr. McKeown, the Rev. Samuel Cutler, the Rev. Mr. Deming, the Rev. Edward Edmunds, and the Rev. W. M. Baker,—men of different evangelical denominations. The lectures for 1875-76 continued eight months, and closed, with the forty-fifth of the course, on the last Monday in May, in Bromfield-street Church.

In October, 1876, the lectures were resumed in the Meionaon; but the hall was found to be too small for the audience. It was, therefore, soon transferred to Park-street Church. Two lectures were given in this large auditorium, when it was found to be much too small, and the audiences were crowded out into Tremont Temple. The first lecture there was given Nov. 13, 1876. This hall will contain from twenty-five hundred to three thousand people, and was often more than full in the winter of 1876–77. During the delivery of a course of thirteen lectures on "Biology," and of eleven on "Transcendentalism," and of eleven on "Orthodoxy," it was often necessary to turn hearers away, as they could not obtain standing-room. From the forty-fifth lecture "The Boston Daily Advertiser" published full stenographic reports of the lectures. The reporter's manuscript was revised by the lecturer. "The New-York Independent" regularly republished the lectures from February, 1876. "The Cincinnati Gazette" did the same; and a large number of newspapers throughout the country published extracts from them. In the course of the winter a few replies to certain statements in the lectures were made by Rev. Dr. James Freeman Clarke and other Unitarians, by Rev. Dr. Miner and other Universalist ministers.

From February, 1876, most of the Boston Monday Lectures were republished in London by the firm of R. D. Dickinson, Farringdon Street. Individual lectures were republished in "The Christian World Pulpit," and other theological serials of Great Britain. At the close of the course for 1876–77, in May, eighty lectures had been given, of which all from the forty-fifth had been published. In September, 1877, James R. Osgood and Company issued "Biology, with Preludes on Current Events," a collection of thirteen Boston Monday Lectures. This volume, at the beginning of December, 1877, was in its twelfth edition. In November the same house issued another course of Mr. Cook's lectures, entitled "Transcendentalism," and announced still another course, entitled "Orthodoxy."

Oct. 1, a course of ten lectures on "Conscience" was opened, and, Dec. 10, a course of ten on "Hereditary Descent." Full stenographic reports, revised by Mr. Cook, are now published in "The Boston Daily Advertiser," "The New-York Independent," "The Cincinnati Gazette," and "The New-York Advocate." Very numerous other papers publish large extracts from them. At least a hundred thousand copies appear weekly. The lectures are regularly republished in London.

It ought to be added, that since the close of his lectures in May, 1877, Mr. Cook has delivered several of them in New-York city, Rochester and Syracuse, N.Y., Princeton, N.J., and various other places; has also supplied various pulpits in Boston and other cities. Before a critic passes any severe criticism on these lectures, he may wisely ask himself whether, without having a previously established reputation, he would be able for two years to interest congregations containing sometimes fifteen hundred hearers, of whom sometimes five hundred are liberally educated men, assembled in the midst of pressing engagements, and in the whirl of a great city; and whether, in addition to his Monday-noon exercises, he would be able to superintend the printing of three volumes of his lectures on abstruse and complicated themes, to preach frequently on the sabbath, and occasionally to deliver sermons, each one of which is from one to two hours in length.

BIOLOGY.
WITH PRELUDES ON CURRENT EVENTS.
Three Colored Illustrations. 12mo. $1.50.

CONTENTS.
LECTURES.
I. HUXLEY AND TYNDALL ON EVOLUTION.
II. THE CONCESSIONS OF EVOLUTIONISTS.
III. THE CONCESSIONS OF EVOLUTIONISTS.
IV. THE MICROSCOPE AND MATERIALISM.
V. LOTZE, BEALE, AND HUXLEY ON LIVING TISSUES.
VI. LIFE OR MECHANISM — WHICH?
VII. DOES DEATH END ALL? INVOLUTION AND EVOLUTION.
VIII. DOES DEATH END ALL? THE NERVES AND THE SOUL.
IX. DOES DEATH END ALL? IS INSTINCT IMMORTAL?
X. DOES DEATH END ALL? BAIN'S MATERIALISM.
XI. AUTOMATIC AND INFLUENTIAL NERVES.
XII. EMERSON'S VIEWS ON IMMORTALITY.
XIII. ULRICI ON THE SPIRITUAL BODY.

PRELUDES.
I. GIFT-ENTERPRISES IN POLITICS.
II. SAFE POPULAR FREEDOM.
III. DANIEL WEBSTER'S DEATH.
IV. CIVIL-SERVICE REFORM.
V. AUTHORITIES ON BIOLOGY.
VI. BOSTON AND EDINBURGH.
VII. THE GULF-CURRENT IN HISTORY.

TRANSCENDENTALISM.
WITH PRELUDES ON CURRENT EVENTS.
12mo. $1.50.

CONTENTS.
LECTURES.
I. INTUITION, INSTINCT, EXPERIMENT, SYLLOGISM, AS TESTS OF TRUTH.
II. TRANSCENDENTALISM IN NEW ENGLAND.
III. THEODORE PARKER'S ABSOLUTE RELIGION.
IV. CARICATURED DEFINITIONS IN RELIGIOUS SCIENCE.
V. THEODORE PARKER ON THE GUILT OF SIN.

VI. FINAL PERMANENCE OF MORAL CHARACTER.
VII. CAN A PERFECT BEING PERMIT EVIL?
VIII. THE RELIGION REQUIRED BY THE NATURE OF THINGS.
IX. THEODORE PARKER ON COMMUNION WITH GOD AS PERSONAL.
X. THE TRINITY AND TRITHEISM.
XI. FRAGMENTARINESS OF OUTLOOK UPON THE DIVINE NATURE.

PRELUDES.

I. THE CHILDREN OF THE PERISHING POOR.
II. THE FAILURE OF STRAUSS'S MYTHICAL THEORY.
III. CHALMERS'S REMEDY FOR THE EVILS OF CITIES.
IV. MEXICANIZED POLITICS.
V. YALE, HARVARD, AND BOSTON.
VI. THE RIGHT DIRECTION OF THE RELIGIOUSLY IRRESOLUTE.
VII. RELIGIOUS CONVERSATION.
VIII. GEORGE WHITEFIELD IN BOSTON.
IX. CIRCE'S CUP IN CITIES.
X. CIVIL-SERVICE REFORM.
XI. PLYMOUTH ROCK AS THE CORNER-STONE OF A FACTORY.

CRITICAL ESTIMATES (AMERICAN).

Rev. Prof. A. P. Peabody of Harvard University, in The Independent.

Joseph Cook is a phenomenon to be accounted for. No other American orator has done what he has done, or any thing like it, and, prior to the experiment, no voice would have been bold enough to predict its success.

We reviewed Mr. Cook's "Lectures on Biology" with unqualified praise. In the present volume we find tokens of the same genius, the same intensity of feeling, the same lightning flashes of impassioned eloquence, the same viselike hold on the rapt attention and absorbing interest of his hearers and readers. We are sure that we are unbiassed by the change of subject; for, though we dissent from some of the dogmas which the author recognizes in passing, there is hardly one of his consecutive trains of thought in which we are not in harmony with him, or one of his skirmishes in which our sympathies are not wholly on his side.

Rev. Dr. Thomas Hill, Ex-President of Harvard University, in the Christian Register.

The attempt of sundry critics to depreciate Mr. Cook's science, because he is a minister, is very ill judged. These Lectures are crowded so full of knowledge, of thought, of argument, illumined with such passages of eloquence and power, spiced so frequently with deep-cutting though good-natured irony, that I could make no abstract from them, without utterly mutilating them.

The Princeton Review.

Mr. Cook has already become famous; and these Lectures are among the chief works that have, and we may say justly, made him so. Their celebrity is due partly to the place and circumstances of their delivery, but still more to their inherent power, without which no adventitious aids could have lifted them into the deserved prominence they have attained. . . . Mr. Cook is a great master of analysis. . . . The Lecture on the Atonement is generally just, able, and unanswerable. . . . We think, on the whole, that Mr. Cook shows singular justness of view in his manner of treating the most difficult and perplexing themes, for example, God in Natural Law, and the Trinity.

Springfield Republican.

This new preacher of modern Orthodoxy delivered his Fifty-first Monday Lecture under the caption " Life or Mechanism — Which ? " this week in the Boston Park-street Church, which was crowded — even to the galleries, aisles, and pulpit-stairs — with an audience mostly composed of men, and representing, to a large degree, the culture and intellect of Boston and vicinity. This Monday Lectureship is now an established institution, and in its growing popularity will tax pretty severely the quality of Mr. Cook. He has so far, however, met the issue squarely, and shows no signs of emptiness or flagging. . . . Mr. Cook has in his favor a happy combination of personal advantages, — a good presence, mental grasp considerable personal magnetism, logical alertness and acuteness; a habit of minute and precise analysis, with sufficient repetition of important details; a poetic and dramatic gift, lighting up what might else be dry and heavy with frequent flashes of wit and fancy, and literary and historical illustration; a restless fervor, the outcome of an excess of physical nervousness, which, however, is never disconcerted; and withal, a fine mastery of good, copious Saxon English.

Boston Daily Advertiser.

At high noon on Monday, Tremont Temple was packed to suffocation and overflowing, although five thousand people were in the Tabernacle at the same hour. The Temple audience consisted chiefly of men, and was of distinguished quality, containing hundreds of persons well known in the learned professions. Wendell Phillips, Edward Everett Hale, Bronson Alcott, and many other citizens of eminence, sat on the platform. No better proof than the character of the audience could have been desired to show that Mr. Cook's popularity as a lecturer is not confined to the evangelical denominations. (Feb. 7.)

It is not often that Boston people honor a public lecturer so much as to crowd to hear him at the noontide of a week-day; and when it does this month after month, the fact is proof positive that his subject is one of engrossing interest. Mr. Cook, perhaps more than any gentleman in the lecture-field the past few years, has been so honored. (Feb. 14.)

The Independent.

We know of no man that is doing more to-day to show the reasonableness of Christianity, and the unreasonableness of unbelief; nor do we know of any one who is doing it with such admirable tolerance, yet dramatic intensity.

George M. Beard, M.D., in the New-York Graphic.

It is said that Mr. Cook misrepresents modern science. This criticism is made mostly by those who do not read all his books; or judge by the original reports at the beginning of the series, or by floating fragments in the papers, or by general hearsay; or very likely by those who themselves know little of science, or at least who are not versed on all sides of his subjects. His work, as it now stands, after many and careful revisions, represents fairly the present state of science on the subject of which he treats, — of the very latest and best researches. Indeed inquirers who will read all of his work, and not part of it, and who are sufficiently endowed with the scientific sense to separate the philosophical reasonings from the facts on which the reasonings are based, will find therein the clearest and most compact statements of the theories and difficulties of evolution, of the movements of bioplasm, and of physiological experiments on decapitated animals and on the electrical irritation of the brain, that appear in popular literature.

Professor John McCrady, in The Literary World.

Mr. Cook's Lectures upon Biology have done good service in making known to a Boston audience the researches of such men as Lionel Beale in England, and the thoughts of such men as Hermann Lotze in Germany, besides the admissions and inconsistencies of the practical materialists, and a valuable review of the whole state of the battle by an able and fearless theological observer like himself. The publication of these Lectures cannot fail to be of service to the extra-scientific world in general. The book well presents to outsiders a certain little-known stage of conservative scientific thought, which they cannot reach anywhere else in so accessible and compact a form. Its extremely popular form, though quite disturbing to the nervous equilibrium of a confirmed man of science, is, nevertheless, well fitted for those it aims to inform, — the great free, intelligent, and religious-minded public, who have not had their heads squeezed by specialistic boards and bandages into strange and fantastic models of approved scientific monstrosity; the people, in short, who have not made philosophical Flatheads of themselves for the sake of some narrow mole's track of scientific investigation known as a "specialty."

This *specialism*, indeed, is aiming to destroy all freedom of thought and speech, and, by consequence, all philosophic thought whatsoever, by forbidding every man to express an opinion on any subject save his own specialty. It has all the narrow intolerance of Comte's Positivism; and I, for one, honor Mr. Cook for his courage in taking it by the beard, and defying it. I heartily recommend his book to the careful reading of everybody who has the interest of scientific conservative thought at heart. Such an one will, at the least, rise from its perusal with a conception of the existing state of the great battle between spirit and matter, very different from that which Mr. Huxley, with the voice of a dragon, lays down in his "Physical Basis of Life;" and, instead of "matter and law devouring spirit and spontaneity," he will see how great cause there is for anticipating the opposite result.

Indeed, the progress of science means, to my apprehension, the very opposite of all that Mr. Huxley contends for in that essay. Spirit and spontaneity are slowly indeed, but surely, advancing along a path which will end in their completely devouring matter and law. The reality of the universe will prove to be the spirit; the illusion of it, the matter; while natural law will declare itself nothing more than the self-consistency of untrammelled spontaneity.

Professor Borden P. Bowne of Boston University, in the Sunday Afternoon.

In the chapters on the Theories of Life, these discussions are, in many respects, models of argument; and the descriptions of the facts under discussion are often unrivalled for both scientific exactness, and rhetorical adequacy of language. In the present state of the debate there is no better manual of the argument than the work in hand. The emptiness of the mechanical explanation of life was never more clearly shown.

Appletons' Journal.

It may be said that the distinguishing and striking characteristic of Mr. Cook's work is, that he pours out the treasures of the latest German thought before audiences and readers whose ideas of science and philosophy have been moulded almost exclusively by that English school, which, as Taine says, tends naturally (by racial inheritance) to materialistic views of life. Our knowledge of the author is confined to what we can obtain from his book; but this is amply sufficient to show that his intellectual equipment has been obtained in Germany, and is truly German in its comprehensiveness and precision. . . . Aside from the rhetorical brilliancy of his style, and the aptness and fertility of his illustrations, Mr. Cook's method of exposition is remarkably effective. By numbering his propositions, and stating them in the concisest possible phrase, he secures a clearness and intelligibility that are seldom so well maintained in a long and complicated argument; and the epigrammatic guise in which most of his principles and conclusions are presented impresses them with peculiar vividness upon the mind of the reader or hearer.

The Eclectic Magazine.

Mr. Cook's rhetorical and literary skill would obtain him a hearing on any subject he chose to discuss; but it is very soon seen, that, beneath the glowing and almost too fervidly eloquent language, there is a force of logic, a breadth of intellectual culture, and a mastery of all the issues involved, such as are seldom exhibited by participants on either side in the great controversy between religion and science. It may be said unqualifiedly that the pulpit has never brought such comprehensiveness and precision of knowledge, combined with such logical and literary skill, to the discussion of the questions raised by the supposed tendency of biological discovery.

International Review.

The lecture-form is retained, and the implied comments of the audience, as given by the reporters, are furnished us, — a feature which will strike readers favorably or otherwise, as their ideas are more or less severe on the composition and make-up of a book. For our part, we like this feature.

The Advance (Chicago).

The reasons given for retaining the responses of the audience, applause, &c., seem to us in this case satisfactory. It is frequently as much a matter of significant interest to know how statements were received by such an audience as to know what the one individual said. This Boston Lectureship is altogether unique in the recent history of popular exposition of abstruse themes. One has to go back to the time of Peter Abelard of the University of Paris for a parallel to it.

The Interior (Chicago).

These Lectures are full of hard thought and eloquent expression. They dwell on the profoundest religious themes, and in the most incisive way. The same power of analysis, sharpness and precision of statement, and gorgeous rhetoric, which characterized the volume on "Biology," are conspicuous here. In these two volumes Mr. Cook has given us the most forcible and readable of all modern defences of essential Christian truth against the scientific and philosophic heresies of the day.

The Standard (Chicago).

The incisive, trenchant style of Mr. Cook has, perhaps, no more admirable adaptation and application than to the demolition of the glittering but specious logic of materialistic philosophy. It is a pleasure to the intellect, as well as to the conscience, to follow Mr. Cook in his irresistible iconoclasm among the images of the theorists who substitute evolution for God in the grand process of cosmogony.

Cincinnati Gazette.

It must be admitted by the most captious critic, that Mr. Cook states his positions with wonderful grace and clearness, and that he fortifies what may appear most paradoxical by a remarkable array of illustration and argument.

Boston Traveller.

There is no denying the fact that Mr. Cook is a born orator. As a popular platform speaker, he has few rivals, and, broadly speaking, we might say no superiors.

Boston Journal.

These Discourses relate to the great problems of life most at issue between science and religion. They were received with eager interest when delivered; and, being republished in whole or in part by the American and English papers, they were, in effect, spoken to an audience on both sides of the sea. Mr. Cook's eloquent and picturesque style — which has in it a touch of Emerson and a touch of Carlyle, as well as qualities peculiarly its own — loses little by transference from the platform to the printed page; and, indeed, the latter form of presentation has its advantages, as being more conducive to the calm and leisure which subjects of so much importance require for their adequate consideration.

New-York Christian Intelligencer.

We believe this book ought to stand and will stand among the very first of the Apologies of the last quarter of this century.

The Christian Union.

Mr. Cook is profoundly interested in his themes. Indeed, he never fails to be kindled into enthusiasm by their transcendent importance. He understands the reach of the physiological questions which he discusses, and the philosophical problems which he essays to solve. His mind is penetrating and subtle. He delights in an argument, and is the last man to fear an antagonist. It would not be easy to decide whether he possesses the logical or the imaginative powers in excess.

Illustrated Christian Weekly.

We enjoy the splendor of Mr. Cook's rhetoric and the brilliancy of his imagination, as in reading a poem.

Church Journal (New York).

His style is peculiar. It is clear, abounding in most expressive figures, with perhaps a slight shading of Carlyleism. But we do not now recall a more forcible writer of the day. His blows at Parkerism, Huxleyism, and Darwinism, come down with sledge-hammer force. He is no mere declaimer. He speaks with the authority of a man who has studied and mastered his subject, and who has fairly dissected the fallacies which he so ably exposes.

The Christian at Work.

Mr. Cook has taken his place as one of the ablest controversialists of the day. His logic is remorseless. He lays every thing under tribute, and drives every nail home.

Worcester Spy.

As a thinker he has notable clearness and strength. His style is full of life and vigor; and he has an admirable mastery of the power of expression; but these alone would not sufficiently explain the great success of his Monday Lectures. The true explanation is, that he selected live questions for discussion, after having studied them, and taken pains to understand them thoroughly. He can meet the most perfectly furnished materialistic speculators on their own ground; is familiar with all the outs and ins of their methods of reasoning; and is able to match their knowledge of the studies and discoveries in physical science, which they use in support of the positions they endeavor to maintain.

Hartford Courant.

The volumes containing his metaphysical speculations and scientific treatment of the problem of religion sell like novels. Mr. Cook is not only a master of the art of putting things, but he is a wit. It is wit none the less because it is used for a serious purpose.

Presbyterian Banner.

The folly of materialistic philosophers has only been exceeded by their arrogance; and it is truly refreshing to find their inflated bubbles so completely punctured and dissipated by the keen thrusts of Mr. Cook's unanswerable logic.

The Penn. Monthly (Philadelphia).

His addresses have been well called prose poems. Nothing could seem less poetical to the eye than his numbered paragraphs. They look like a series of theses set up for the defiance of all comers. But ear and sense alike are captivated as we read, and we are forced to recognize a master of English prose.

Religious Herald (Richmond, Va.).

No man in America is just now attracting more attention than Joseph Cook, and his Titanic blows are telling on the materialistic scepticism of the day. . . . He is clear, axiomatic, and irresistible through all his arguments, and, while always courteous to opponents, is often keenly satirical.

The Theological Medium, Nashville, Tenn.

His learning is immense, his faculty of order eminent, his imagination very brilliant, and his logic strong and close.

New Orleans Times.

The Lectures are crowded with eloquent passages, telling satire, and keen, critical, and precise reasoning.

San Francisco Evening Bulletin.

The style is peculiarly vivid, presenting occasionally some of the characteristics of Carlyle. The book, in consequence of its scope and general attractiveness, is destined to become very popular.

San Francisco Bancroft's Messenger.

Possessed of a calm, critical, and methodical mind, Mr. Cook has constructed, from the material at his disposal, about a dozen of the most interesting essays that have yet appeared on the relation of religion and science. On almost every page of the volume, eloquence leaves its mark.

San Francisco Evening Post.

Emotion, clearness, and sound sense are the weapons with which he produces conviction.

CRITICAL ESTIMATES (FOREIGN).

Rev. R. Payne Smith, Dean of Canterbury.

The lectures are remarkably eloquent, vigorous, and powerful, and no one could read them without great benefit. They deal with very important questions, and are a valuable contribution towards solving many of the difficulties which at this time trouble many minds.

Rev. Dr. Angus, the College, Regent's Park.

These Lectures discuss some of the most vital questions of theology, and examine the views or writings of Emerson, Theodore Parker, and others. They are creating a great sensation in Boston, where they have been delivered, and are wonderful specimens of shrewd, clear, and vigorous thinking. They are, moreover, largely illustrative, and have a fine vein of poetry running through them. The Lectures on the Trinity are capitally written; and, though we are not prepared to accept all Mr. Cook's statements, the Lectures, as a whole, are admirable. A dozen such lectures have not been published for many a day.

Rev. Alexander Raleigh, D.D., of London.

The Lectures are in every way of a high order. They are profound and yet clear, extremely forcible in some of their parts, yet, I think, always fair, and as full of sympathy with what is properly and purely human as of reverence for what is undoubtedly divine.

Rev. John Ker, D.D., of Glasgow.

My conviction is, that they are specially fitted for the time, and likely above all to be useful to thoughtful minds engaged in seeking a footing amid the quicksands of doubt. There is a freshness, a power, and a felt sincerity, in the way in which they deal with the engrossing questions of our time, and, indeed, of all time, which should commend them to earnest spirits which feel that there must be a God and a soul, and some way of bringing them together, and which yet have got confused amid the negations of the dogmatic scepticism of our day. I had the pleasure of meeting Mr. Cook four years ago, when he was visiting Europe to make himself acquainted with different forms of thought, and I could see in him a power and resolution which foretold the mark he is now making on public opinion.

Rev. C. H. Spurgeon.

These are very wonderful Lectures. We bless God for raising up such a champion for his truth as Joseph Cook. Few could hunt down Theodore Parker, and all that race of misbelievers, as Mr. Cook has done. He has strong convictions, the courage of his convictions, and force to support his courage. In reasoning, the infidel party have here met their match. We know of no other man one-half so well qualified for the peculiar service of exploding the pretensions of modern science as this great preacher in whom Boston is rejoicing. Some men shrink from this spiritual wild-boar hunting; but Mr. Cook is as happy in it as he is expert. May his arm be strengthened by the Lord of hosts!

The Methodist (London).

Let all who need inspiring afresh with love for the holy gospel, and a new and manly ardor for its proclamation, read the Monday Lectures of the Rev. Joseph Cook.

The Lucknow Witness (India).

The paper will be greatly improved next year, and will contain several new and interesting features : among them we may mention the publication of the Lectures of the Rev. Joseph Cook, now making such a stir in Europe and America.

London Quarterly Review.

For searching philosophical analysis, for keen and merciless logic, for dogmatic assertion of eternal truth in the august name of science such as thrills the soul to its foundations, for widely diversified and most apt illustrations drawn from a wide field of reading and observation, for true poetic feeling, for a pathos without any mixture of sentimentality, for candor, for moral elevation, and for noble loyalty to those great Christian verities which the author affirms and vindicates, these wonderful Lectures stand forth alone amidst the contemporary literature of the class to which they belong.

⁎⁎⁎ For sale by all booksellers. Sent, postpaid, on receipt of the price, by the Publishers,

HOUGHTON, OSGOOD & CO., Boston.

www.ingramcontent.com/pod-product-compliance
Lightning Source LLC
Chambersburg PA
CBHW031247250426
43672CB00029BA/1369